Earth Science

God's World, Our Home

A Mastery-Oriented Curriculum

NOVARE
SCIENCE & MATH

Austin, Texas
2015

Earth Science

God's World, Our Home

A Mastery-Oriented Curriculum

Kevin Nelstead

NOVARE
SCIENCE & MATH

Austin, Texas
2015

Published by

Novare Science & Math
P. O. Box 92934
Austin, Texas 78709-2934
novarescienceandmath.com

NOVARE
SCIENCE & MATH

Printed in the United States of America

ISBN: 978-0-9863529-8-0 (This ISBN applies only to this preprint edition of the text.)

Cover design by Julie Kennedy, Digital City Designs, www.DigitalCityDesigns.com.

For the complete catalog of textbooks and resources available from Novare Science and Math, visit novarescienceandmath.com.

PREPRINT NOTICE

This volume is a PREPRINT of a portion (the first 9 chapters) of the forthcoming title *Earth Science: God's World, Our Home*, by Kevin Nelstead, published by Novare Science & Math. Content in Chapters 1–8 has been reviewed by our team of expert scientists for accuracy. Review of Chapter 9, and final editing and review of all the chapters has not been completed.

Acknowledgements

We wish to express our appreciation to the author of this text, Kevin Nelstead, who has contributed such an excellent treatment to our textbook line-up.

We are indebted to those who reviewed this text during its development: Chris Mack, PhD; Steven Mittwede, PhD, and Ron DeHaas, and deeply appreciate their work. Thanks also to our excellent copy editor Emily Cook.

The Author

Kevin Nelstead earned a Bachelor of Science degree in Earth Science from Montana State University, and a Master of Science degree in Geology from Washington State University. His Master's degree research involved the study of Quaternary (ice age) volcanic ash deposits in the Channeled Scablands and Palouse Hills area of eastern Washington.

Kevin has seventeen years of experience working as a Senior Cartographer, Geospatial Analyst, and Natural Resources Specialist for the United States federal government. Most of this work has involved creation of topographic maps and interpretation of aerial and satellite imagery. His maps and atlases have placed as high as second in national and statewide professional map competitions.

After earning his teaching certificate in Chemistry from the University of Missouri-St. Louis, Kevin taught science in three different Christian schools: a Roman Catholic high school, a classical Christian school, and an international Christian school. Kevin also regularly teaches Geographic Information Systems courses through his work, and taught as an instructor for one year at Montana State University-Billings.

Kevin has been married to his high school sweetheart, Shirley, for over thirty years, and has four adult children. Kevin and Shirley were missionaries in Romania for over five years, serving with ReachGlobal, which is the international ministry of the Evangelical Free Church of America. They are now members of a Presbyterian Church in America church in Montana.

When he has spare time, Kevin writes about Christianity, science, and the environment at GeoChristian.com.

Reviewers

Steven K. Mittwede	Earth Science & Theology Teacher, Providence Classical School, Spring, Texas PhD, University of South Carolina, Geology MS, University of South Carolina, Geology BS, The College of William and Mary in Virginia, Geology EdS candidate, Columbia International University, Educational Leadership MTh, University of Glamorgan/Wales Evangelical School of Theology, Modern Evangelical Theology MA, Columbia International University, Intercultural Studies
Ronald J. DeHaas	Consulting Petroleum Geologist, Michigan Basin and CEO, Covenant Eyes, Inc. PhD Candidate, Geology, University of Michigan MS, Geology, The Ohio State University BS, Geology, The Ohio State University
Chris Mack	Adjunct Faculty, University of Texas at Austin PhD, University of Texas at Austin, Chemical Engineering MS, Electrical Engineering, University of Maryland BS degrees in Physics, Electrical Engineering, Chemistry, Chemical Engineering, Rose-Hulman Institute of Technology

Earth Science

God's World, Our Home

NOVARE
SCIENCE & MATH

Contents

PREPRINT NOTICE

This volume is a PREPRINT of a portion (the first 9 chapters) of the forthcoming title *Earth Science: God's World, Our Home*, by Kevin Nelstead, published by Novare Science & Math. Content in Chapters 1–8 has been reviewed by our team of expert scientists for accuracy. Review of Chapter 9, and final editing and review of all the chapters has not been completed.

Preface

For Teachers

The book you have before you is clearly not a typical middle school earth science text. It is not two inches thick, it does not weigh five pounds, and it does not attempt to cover every topic in all of earth and planetary science. These are not accidents.

We believe that science education, including textbook design, needs a major overhaul. The design philosophy behind this text represents our idea of where to start. Our philosophy of pedagogy—and thus textbook design—is based on three core principles—*Mastery*, *Integration*, and *Kingdom Perspective*. We summarize these principles this way:

Mastery

The norm for classes in contemporary schools is what we call the *Cram–Pass–Forget cycle*. Students cram for tests, pass them, and then forget most of what they crammed in just a few weeks. Teachers across the nation know exactly what this looks like because they see it day after day. This cycle is a waste of time for teachers and students. Instead, students should *learn*, *master*, and *retain* what they have learned. Realizing this in the classroom requires both the teacher and the students to make significant changes in the ways they approach the tasks of teaching, testing, practicing, and studying. For textbooks, a mastery orientation requires that the curriculum cover fewer topics, and cover them deeply. When students engage with content, their efforts should be directed at comprehension, practice, and retention of important concepts, not mere exposure to masses of material they will not remember. Methods that promote the Cram–Pass–Forget cycle must be replaced with more effective methods.

In a mastery-oriented learning environment, students will notice that the workload is demanding, and initially some of them will resist. But all the students will learn far more than they ever have in a course. They will remember the material for years, and most of them—even those who aren't much interested in school—will come to recognize the value of such a learning experience. Some will recognize this immediately, some only after they go off to college and find themselves ahead of their peers because of things they remember from many years before.

For more information about how mastery concepts should be implemented with this text, see the sections below entitled *Companion Book and Resource CD* and *Notes on Using this Text.*

Integration

For a variety of complex reasons, effective science instruction requires a number of content areas to be deeply embedded in the science curriculum—content areas often considered tangential and not represented adequately in curriculum materials. These content areas include scientific process, basic epistemology, mathematics, scientific history, and English language usage. Typically, science classes do not place the necessary emphasis on these areas, and as a result, students fail to have a coherent and effective learning experience. Science teachers need to think deeply about how their courses are addressing the need for these and other key areas of integration and make adjustments to curriculum, teaching methods, assessments, and coordination between science courses.

An example of how integration is realized in this text is that the questions and exercises in every chapter and in every experiment require students to write out answers in complete sentences. Another example is the introduction of graphical (and thus quantitative) methods in both the text and experiments. A third example is the discussion of science and theories in Section 3.1. This discussion is essentially about epistemology—kinds of knowledge and how we know what we know—although the term epistemology is not used in the text. We strongly assert that without a proper epistemology students will not correctly understand the nature of scientific theories or the role they play in the ongoing process of scientific research. Nor will they understand the difference between scientific claims and truth claims. Thus, it is critical for every science text to engage these questions at an age-appropriate level.

Kingdom Perspective

The study of science provides us with unique ways of seeing God's creative presence in the world. Bringing biblical faithfulness to science classes will not be accomplished by simply folding in a few Bible verses or prayers. In fact, much more is involved. Science and math teachers need to think broadly about how we fulfill Christ's mandate to love God with all of our mind, how we teach our students effectively to engage issues, and how we perceive God's fingerprints in creation.

We also need to engage thoughtfully with the scientific claims of our day. It is not a scientific claim to say that the universe got here by itself; that is a metaphysical claim based on an atheistic worldview. But it is a scientific claim to say that the universe began with the Big Bang and is now 13.77 billion years old. The scientific evidence behind this claim is vast, and we believe an appropriate science text is one that teaches students productively to engage such claims. We do not believe it is appropriate to teach students to be dismissive of claims like this one simply because they do not line up with certain ways of interpreting Genesis. We hold this as those who fully believe the Bible, who love reading Genesis and the rest of Scripture, and who accept the strong evidence for an old universe.

Our ideas about all three parts of this core philosophy are described in more detail in John D. Mays' book *Teaching Science so that Students Learn Science.* This book and a more detailed description of our textbook philosophy may be found on our website, novarescienceandmath.com

7th Grade? 8th Grade? Considerations for Science Programming

This text has been specifically developed for application in grades seven or eight. If used in seventh grade, teachers may consider as optional parts of Section 4.1 on Atoms, Elements, and Crystals, especially Section 4.1.4 on Chemical Bonds. For students who study Physical Science in sixth or seventh grade and use this text in eighth grade, Section 4.1 will provide a good review of material covered in the physical science course.

Lab Journals and Lab Reports

The overall goal of experiment documentation for middle school students is to continue laying the foundation for writing full lab reports—from scratch—when they get to high school. The target is for students to begin writing full lab reports in ninth grade, and the standard for such reports is presented in John D. Mays' book *The Student Lab Report Handbook* (available on our website). Toward that end, we believe lab reports at the middle school level should focus on students describing what they did, presenting their results, and engaging with the questions.

Part of integrating English language development into science instruction is accomplished through the use of lab journals. Each of the eight experiments included in this text requires students to document their work in a lab journal. Details about using lab journals are in the Preface for Students, as well as in the first chapter of *The Student Lab Report Handbook.*

In each of the Experimental Investigations, we have posed questions for students to engage with as they consider their results and observations. We have left decisions about the specifics of the student reports to the individual instructor; requirements should be based on the preparedness and background of the students in a given class.

Conducting Experiments

It is very important for all students to conduct the eight experiments included in this text. Interacting with content about the Earth through the text alone is an inadequate approach to this subject. Students need to study the actual topo maps and handle actual minerals and rocks.

The instructions written in the Experimental Investigations provide the information students need. Additional details on resources, materials, suppliers, costs, and ways to enhance the lab activities are included on the Resource CD for this text (available at novarescienceandmath.com).

Companion Book and Resource CD

As mentioned at the beginning of this preface, a mastery-based learning environment entails new teaching methods and new study techniques. These are described in detail in *Teaching Science so that Students Learn Science*, and we encourage teachers to read that book and implement the strategies discussed there. In particular, the discussion in Chapter 7 (Science as a Cumulative Discipline) reviews the philosophy of mastery-based learning and methods necessary to realize it. Two of the most important strategies from that chapter are the weekly cumulative quiz and the Weekly Review Guides.

The resource CD available to accompany this text includes a full year's worth of weekly quizzes, Weekly Review Guides, semester exams, and sample answers to questions. As described in *Teaching Science so that Students Learn Science*, for students at this level we promote the use of a 30-minute, *cumulative* weekly quiz in place of chapter tests. Cumulative quizzes include questions each week from Objectives Lists throughout the portion of the text previously covered, all the way back to the beginning of the course. This assessment method is a thoroughly demonstrated way to realize significant improvements in student retention.

To enable students to succeed at the weekly cumulative quizzes, particularly after the sixth or eighth quiz when the quantity of material to remember becomes significant, teachers must consistently engage students with correct study methods. One of the essential components of the mastery-based program is the Weekly Review Guides. This important tool helps students to know how to study—and what to study—each week when long-term retention is the goal. The Resource CD includes a year's worth of Review Guides, which teachers can begin issuing to students in the third week of the course.

Notes on Using this Text

One of the major motivations behind the design of this text is for students to be delivered from the Cram–Pass–Forget cycle alluded to earlier. Compared to other texts, this book is small. An appropriate amount of content has been assembled for students to master in a single year of study.

But for mastery to be realized, certain practices need to be a regular part of the classroom experience. Regular review, rehearsal, and practice of older material must occur alongside the study of new material. Otherwise, students will simply forget things a few weeks after being tested on them.

Here are a few specific practices that should be regularly present in classes using this text. These practices simply represent good teaching, and should be part of any well-structured course.

1. *Class Discussion* Discuss concepts, ask questions, and give students opportunities to express concepts in their own words. Use the "Learning Check" questions at the end of each section and the Chapter Exercises as questions to stimulate discussion. All of these questions should be discussed in class. Take time with this type of discussion and don't hurry. This is why the book is mod-

estly sized—so there will be adequate time for extended discussion and deep engagement with the material.

2. *Group Study* Divide students into groups for teamwork on answering some of the more difficult questions in the Chapter Exercises. Then have teams take turns reporting their answers to the class. Stimulate discussion between teams about the merits of different answers to a particular question, and seek a class consensus on a well-formed answer.

3. *Review Discussion* Go through questions again from past chapters. It is also fun to create competitions among teams in which teams score points by giving good answers to review questions. Some kind of activity like this should occur at least once every 3–4 weeks throughout the duration of the course.

4. *Enrichment Activities* Although concepts are covered thoroughly in the text, understanding will be enhanced and memory will be strengthened when students engage with the content in activities outside the text. Classes should incorporate full-length instructional videos, classroom samples of different rocks and minerals, internet images, You Tube videos, games, student presentations, projects, nature field trips, and other activities. Teachers should emphasize activities designed to make students *active* participants in the learning process, rather than simply passive observers.

5. *Learning Rock Names* The vocabulary list for Chapter 5 can appear daunting. Encourage your students to study these terms—as well as the terms from all other chapter vocabulary lists—using flashcards and frequent review. For challenging lists of terms such as the Chapter 5 vocabulary list, teachers should stretch out the learning process over an extended period of time. Mix class discussions with frequent use of classroom samples of the specific types of rocks listed in the vocabulary list, and use games and competition to make learning remembering the terms exciting and rewarding.

Preface

For Students

This world belongs to God, our Father and Creator. As the 24th Psalm states: "The earth is the Lord's and the fullness thereof, the world and all those who dwell therein, for he has founded it upon the seas and established it upon the rivers." God's world is a wonderful and precious gift to us, for it is our home. Adam was placed in the garden of Eden "to work it and keep it." This biblical language (*work* and *keep*) means that Adam was to *develop* the garden so that it would be fruitful, and *preserve* the garden so that it would continue always to provide a suitable home for human beings and all the other creatures. This charge to Adam from God is often called the *Cultural Mandate*—a mandate from God for the human race wisely to develop and care for all the gifts we have received, including our minds, skills, communities, technologies, and, of course, the beautiful world he made. As hymnist Maltbie Babcock wrote in 1901,

> This is my Father's world, and to my listening ears
> All nature sings, and round me rings the music of the spheres.
> This is my Father's world: I rest me in the thought
> Of rocks and trees, of skies and seas;
> His hand the wonders wrought.

Stewarding God's Gifts

God's world—our home—is fascinating and fun to learn about. We have had a lot of fun preparing this text, and we hope you have a lot of fun using it as part of your study of Earth Science. One of our goals for you is that you develop a *love* for the Earth. We want you to appreciate its beauty and the delicate way the systems and organisms in this world are balanced. The intricate balance and beauty we see in nature are breathtaking!

It is important for us to act as good *stewards* of the gifts God has given to us. The Earth is certainly a great gift, and in our time the Earth is in need of our care and attention. But we humans generally do not properly care for things unless we love them, and we generally don't love things unless we spend time with them. We encourage you to find ways to spend time with the Earth—not just riding in cars around town, but exploring streams, walking in the woods, playing in the ocean, and hiking in the mountains. If you are going on vacation with your family, we encourage you to suggest to your parents to include stops at National Parks, National Forests, or other natural spots in the trip plan so that you can experience the glories of this beautiful planet for yourself.

As you develop your understanding of Earth's processes, we hope you will also understand that our lovely planet is under a great deal of stress right now. The health of our environment and the millions of creatures that inhabit it cannot be sustained unless people everywhere work together to develop new habits of land management, new attitudes toward recycling and conservation, and new energy technologies. As a young person, you will soon be entering adulthood and making decisions about how to direct your energies as a servant in God's Kingdom. We hope you will make your decisions with godly stewardship of the Earth well in mind.

Mastery

Another of our goals for you in this course is mastery: we don't want you simply to cram for tests, pass them, and then about three weeks later forget what you crammed. Instead, we want you to *learn*, *master*, and *retain* what you learn about the beautiful planet God has made for us to inhabit. For this reason, the weekly quizzes we have made available to your instructor always include questions about things you studied in previous chapters. This happens all year long, and your quizzes when you are in Chapter 14 will still be asking you questions from Chapters 2, 4, 6, and all the other chapters.

To succeed at the weekly quizzes, you need to study in particular ways. First, we suggest that you make flashcards for the vocabulary terms in each chapter and that you review these flashcards regularly. Second, we suggest that you make use of the Weekly Review Guides that are available to your instructor. Go through the review activities carefully each week so that you can remember the material from early chapters. Third, you should also study the Objectives List at the beginning of each chapter and make sure you can perform every item on the list.

Learning any subject is frustrating if you feel like you are jumping through hoops but not accomplishing anything that is valuable or lasting. But learning is *a lot* more fun when you feel like you are truly learning, remembering, and growing. The study and review techniques listed above, and others that your instructor will introduce to you, will make the study of Earth Science a wonderful experience.

Who knows? You may even find that you love this subject so much that you want to become a geologist, climate modeler, or environmental researcher!

Lab Journals

A third goal for our course is that you will expand your English language skills so that you can express yourself clearly, and write accurate answers to scientific questions. One way to practice using English in scientific study is carefully to document your experimental work in your own lab journal. Here is some advice about using a lab journal with the Experimental Investigations in this text:

1. Keep your lab journal very neat and well organized. Don't doodle in it or mess it up.
2. Don't use a spiral notebook. Use a bound composition book with quadrille (graph) paper. (Quadrille paper makes it easy to set up tables and graphs.) A popular one to use is the Mead 09127, available at office supply stores and pictured to the right.
3. Put your name on it in case you misplace it.
4. For every experiment you work on, enter the following information:
   - the date (always enter the date again every day you work, as one always does with a journal)
   - the names of team members working with you (enter these also every day you work, so you have a record of who is there and who is not each time you meet)
   - a list of all equipment, apparatus, materials, and supplies you use in conducting the experiment
   - tables with *all* of your data, with the original units of measure
   - calculations or unit conversions you perform as part of the experiment
   - observations or notes about anything that happens that you may need to write about in your report or remember later, including records of work that must be repeated and why
   - methods or procedures you use, and the reasons for using them

There are other items to enter in your journal that will become more important as you get into high school and college (such as sources, contacts, and prices for special chemicals or parts you have to order), but the list above should cover the things that you need to worry about for now. Take pride in maintaining a thorough lab journal. Make it your habit always to have it with you when you work on your experiments, and always to document your work in it.

NOVARE
SCIENCE & MATH

Earth Science

God's World, Our Home

A Mastery-Oriented Curriculum

Chapter 1

Earth In Space

In December 1968, three astronauts went into orbit around the moon in the Apollo 8 spacecraft. This was the first time that humans had gone further from Earth than just a few hundred kilometers above the surface. As they circled the moon, the Apollo 8 crew were the first people to have a direct, close-up view of the desolate craters, plains, and mountains of another world. Since radio signals cannot go through or around the solid sphere of the moon, they were out of radio contact with Earth for about thirty minutes each time they went behind the moon. As they came back from their fourth journey behind the moon, they saw something that caught their attention far more than the unexplored surface of the moon—they saw Earth rising above the lunar horizon. The moon was gray and barren; Earth on the other hand hung brilliantly in the sky with its blue oceans, white clouds and ice caps; and variously-hued colors of the continents. They journeyed 384,000 kilometers to the moon, but then realized that the most important object in their view was not the moon, but Earth.

Objectives

After studying this chapter and completing the exercises, you should be able to do each of the following tasks, using supporting terms and principles as necessary.

1. Define and describe the four major Earth systems.
2. Give examples of specialties practiced by different kinds of Earth scientists.
3. Describe Earth's location in the solar system, galaxy, and universe.
4. Define what a habitable zone is, and apply this definition to Earth's place in the solar system, and the solar system's place in the Milky Way galaxy.
5. Give examples of ways in which Earth seems to be "just right" for complex life.
6. Describe Earth's orbit around the sun.
7. Explain how the tilt of Earth's axis causes the seasons.
8. Explain why it is hot in the summer and cold in the winter in the Northern Hemisphere.
9. List in order the phases of the moon throughout the lunar cycle.
10. Describe the position of the sun, Earth, and moon for each phase of the lunar cycle.
11. Explain what causes partial and total solar eclipses, and explain why they do not occur every month.
12. Explain what causes lunar eclipses.
13. Explain how solar, lunar, and lunisolar calendars work, and give an example of each.
14. Compare the Julian and Gregorian calendars, explaining how the Gregorian calendar corrected a flaw in the Julian calendar.

Vocabulary Terms

You should be able to define or describe each of these terms in a complete sentence or paragraph.

1. lithosphere
2. hydrosphere
3. atmosphere
4. biosphere
5. geology
6. oceanography
7. meteorology
8. geocentric model
9. heliocentric model
10. habitable zone
11. galaxy
12. Milky Way galaxy
13. universe
14. perihelion
15. aphelion
16. ecliptic
17. summer solstice
18. winter solstice
19. vernal equinox
20. autumnal equinox
21. Tropic of Cancer
22. Tropic of Capricorn
23. Arctic Circle
24. Antarctic Circle
25. lunar phase
26. new moon
27. waxing crescent
28. first quarter
29. waxing gibbous
30. full moon
31. waning gibbous
32. last quarter
33. waning crescent
34. solar eclipse
35. partial solar eclipse
36. total solar eclipse
37. umbra
38. penumbra
39. total lunar eclipse
40. lunisolar calendar
41. lunar calendar
42. solar calendar
43. Julian calendar
44. Gregorian calendar

1.1 An Introduction to Earth Science

The image on the opening page of this chapter is a picture taken by astronauts on the spacecraft Apollo 17, the final mission to the moon in 1972. From space, one can see that Earth is not a monotonous place. Some of its land surface is brown and barren, and other parts are covered with lush vegetation. Greater than 70% of the surface is covered by oceans. Both land and the oceans near the poles are covered with water in a different form: snow and ice. Forming a thin layer on top of all of these is the atmosphere, with its ever-changing patterns of clouds.

1.1.1 Earth Systems

What you see from space can be divided into different Earth systems, illustrated in Figures 1.1 through 1.5. Scientists think of these systems as concentric spheres, with the solid Earth at the center, then water, then the air; with living organisms present in all three. This gives us four primary Earth systems: the lithosphere, the hydrosphere, the atmosphere, and the biosphere.

The *lithosphere* is the rigid outer layer of Earth, composed mostly of solid rock. The lithosphere includes Earth's crust and the upper part of the underlying mantle. The rocks of the crust are exposed in many places at the surface of Earth. There are two basic types of crust: continental crust and oceanic crust. The continental crust is composed largely of a lighter-colored rock called granite, and averages about 30 to 40 km (20–30 mi) in thickness. The oceanic crust, on the other hand, is typically about 5 km (3 mi) thick, and is composed of a dark, dense igneous rock (that is, formed from molten rock) called basalt. Beneath the lithosphere is the rest of Earth's rocky mantle, and beneath the mantle is Earth's iron core. You will learn more about Earth's crust, mantle, and core in Chapter 7.

The *hydrosphere* is the part of Earth that is made out of water. This water is present as a liquid in the oceans, seas, lakes, rivers, and as groundwater in pores in rocks and soil. Additionally, water is present as a solid—snow and ice—on land and the polar oceans, and as a gas in the atmosphere. Most of Earth's water is present as salt water in the oceans. The greatest amount of fresh water—that is, non-salty water—is contained in ice, primarily in the ice caps that cover Greenland and Antarctica. Compared to the oceans and ice caps, there is only a tiny amount of water in lakes and streams.

Figure 1.1. Lithosphere—the solid, rigid outer layer of Earth. This quarry is in Australia.

Earth is surrounded by a layer of gases called the *atmosphere*. The thickness of the atmosphere compared to the rest of the planet is like the peel of an apple compared to the rest of the fruit. The most abundant gas in the atmosphere is nitrogen (78%), followed by oxygen (21%). The remaining 1% is made up of argon, carbon dioxide, and a number of gases that are present in small proportions. The atmosphere also contains a variable amount of water vapor—water in its gaseous state.

Figure 1.2. Hydrosphere—Earth's water. Havasu Falls is in the Grand Canyon in Arizona.

The atmosphere serves a number of functions. The oxygen in the atmosphere is necessary for respiration for most living things, and the carbon dioxide is necessary for photosynthesis. The atmosphere also helps to maintain the temperature of the surface of Earth in a range that is suitable for advanced organisms such as plants and animals. In addition, the gases of the atmosphere help prevent various types of harmful radiation from the sun and deep space from reaching Earth's surface.

The *biosphere* is made up of all organisms that live on Earth along with the environments in which they live. Life exists in almost every environment on Earth:

Figure 1.3. Atmosphere—The gaseous layer that surrounds Earth. This thunderstorm occurred over New Mexico.

Figure 1.4. Biosphere—All living things on Earth. This temperate rainforest is in Redwood National Park in California.

Figure 1.5. All four Earth systems—lithosphere, hydrosphere, atmosphere, and biosphere—are interacting along this Atlantic Ocean coastline in Maine.

the surface, the soil, the air, the deep sea, hot springs, ice, and even cracks in hot rocks thousands of meters beneath the surface.

These four systems all interact with each other. It is obvious that organisms in the biosphere are dependent on the lithosphere, hydrosphere, and atmosphere for oxygen, water, nutrients, and space to live. However, not only does the biosphere use resources from the other systems, the biosphere in turn affects those systems as well. Plants have changed the atmosphere by producing oxygen. Plants also help to break down minerals in the soil by removing nutrients and exchanging water with the hydrosphere. Likewise, there are interactions between the lithosphere, hydrosphere, and atmosphere. For example, water—coming from the atmosphere—falls on Earth and causes erosion of the soil, which is part of the lithosphere. Many of the interactions between the solid Earth, water, air, and life are extraordinarily complex and are not fully understood.

1.1.2 Subdivisions of Earth Science

Figure 1.6. A geologist sampling 1150°C (2100°F) lava at Kilauea, a volcano in Hawaii.

Earth scientists work in a number of specialties. Traditionally, these are broken down into three major subdivisions: geology, oceanography, and meteorology, illustrated in Figures 1.6 through 1.8.

Geology is the study of the materials that make up Earth and the processes that change Earth over time. Geologists, of course, study rocks, but they also study a range of materials and processes that will be covered in this book, such as volcanoes, Earthquakes, fossils, streams, glaciers, water resources, mineral resources, and energy resources.

Figure 1.7. Oceanographers can use submersibles such as DeepWorker to explore the ocean's depths.

Figure 1.8. Meteorologists using radar to study a tornado.

Oceanography is the study of the oceans. An oceanographer might study ocean currents, processes that occur on beaches, mineral resources on or beneath the ocean floor, or organisms that live in the ocean.

Meteorology is the study of the atmosphere. Meteorologists not only make weather observations and forecasts, they are interested in studying air pollution and understanding long-term trends and changes in climate at various places.

1.1.3 Further Specializations

Due to the explosive growth of scientific knowledge, it is impossible for a geologist, oceanographer, or meteorologist to have complete knowledge of his or her subject. Usually Earth scientists have a broad knowledge of the sciences, but then have a specialty which is the focus of their work. Some specialties of Earth sciences include:

- Climatology — The study of climate, which is the long-term average of weather conditions in an area.
- Ecology — The study of the interactions between organisms and their environments. Ecology is often considered to be a topic in biology, but it is also an important topic in Earth sciences. Organisms affect Earth, and Earth affects the organisms that live on it.
- Geochemistry — The use of chemistry to understand Earth processes.
- Geophysics — The use of physics to understand Earth, including its shape, interior, magnetism, and surrounding space.
- Hydrology — The study of the movements and quality of water, either on Earth's surface or under it.
- Marine biology — The study of life and ecosystems in the oceans.
- Mineralogy — The study of the formation, composition, and distribution of minerals.
- Paleontology — The study of past life on Earth and how it has changed over time.

- Petroleum geology — The study of the location, migration, and production of oil and gas resources.
- Petrology — The study of rocks.
- Planetary geology — Application of geological principles to other worlds, such as planets, moons, and asteroids.
- Volcanology — The study of volcanoes.

This list is just a fraction of the specializations within Earth sciences. Even within these specialties, a scientist usually focuses on an even narrower topic. A climatologist might be most interested in desert climates, a paleontologist might specialize in coral fossils of the Jurassic Period, or a volcanologist might focus on the chemical composition of volcanic rocks produced from volcanoes like the ones in Hawaii. Even within these specializations, however, the best scientists will be those who can relate their data to the work being done by workers in other specializations. Because of this, scientists often work in teams and attend meetings with other scientists so they can exchange ideas and look for interactions and relationships between their work and that of others.

Learning Check 1.1

1. Distinguish between the lithosphere, hydrosphere, atmosphere, and biosphere.
2. Suggest two ways that the biosphere interacts with each of the other Earth systems.
3. Give a definition for each of the three major subdivisions of Earth science that you will be learning about in this course.

1.2 Earth in the Solar System, Galaxy, and Universe

We cannot completely understand Earth without having an understanding of Earth's place in the solar system, galaxy, and universe. After all, things that happen at a great distance from Earth can greatly influence our planet. Energy from the sun, produced by nuclear fusion of hydrogen and helium in the sun's core, constantly bathes Earth's land, water, and air in the form of electromagnetic radiation. There is gravitational attraction between Earth and the sun, moon, and other planets in the solar system. Rare astronomic events can influence Earth as well, such as the collision of large meteorites or even asteroids with Earth.

1.2.1 Earth in the Solar System

Until the sixteenth and seventeenth century, most scientists believed that Earth was at the center of the physical universe. It was thought that the sun, moon, and five planets that were known at that time (Mercury, Venus, Mars, Jupiter, and Saturn) all orbited around Earth in perfectly circular orbits, as shown in Figure 1.9. In this model, the stars were points of light that also revolved around Earth. This

Earth-centered picture of the universe is known as a *geocentric* model. The story of how scientists changed their minds about this model of the universe is fascinating, and was an important turning point in the history of science, but it is a topic for another course. It took about one hundred years from the work of Nicolaus Copernicus (1473–1543) until after the death of Galileo (1564–1642) for most scientists to abandon the geocentric model of the universe.

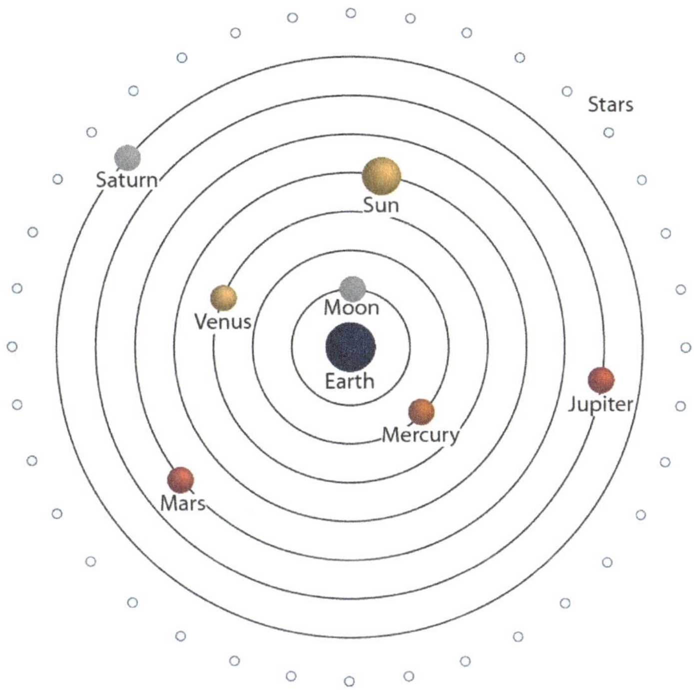

Figure 1.9. In the geocentric model, Earth is at the center of the universe, and all other bodies orbit Earth.

Today we have a very different picture of the place of Earth in the solar system and of the universe as a whole, illustrated in Figure 1.10. We now understand that the sun, not Earth,

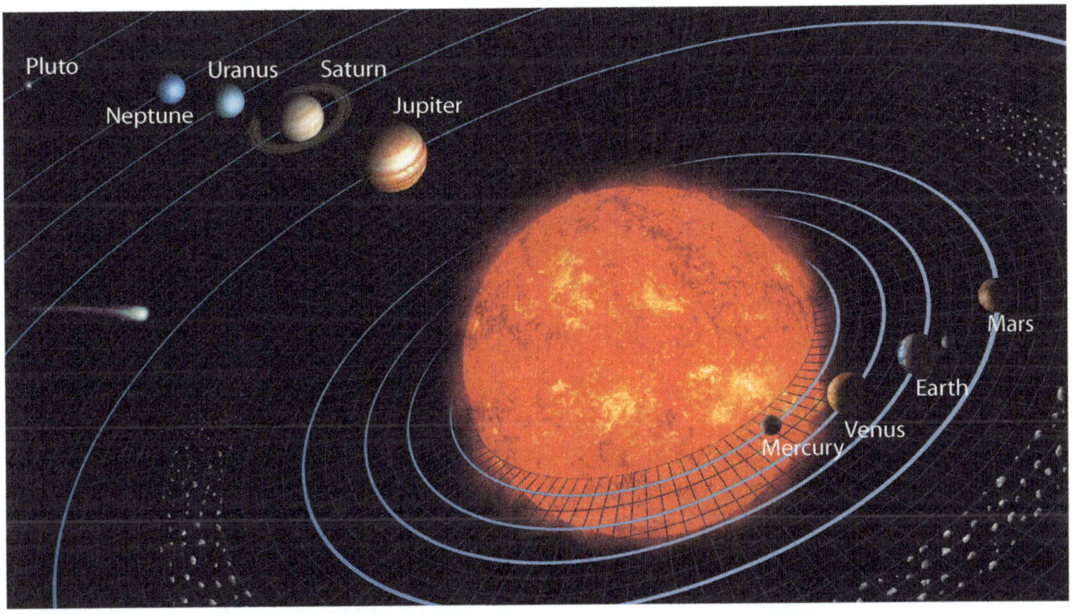

Figure 1.10. In the heliocentric model, planets, asteroids, comets, and other solar system bodies orbit the sun. Our sun is just one of over 100 billion stars in the Milky Way galaxy.

is at the center of the solar system. This model of the solar system is known as a *heliocentric* model. Earth is one of eight planets that orbit the sun. The four innermost planets—Mercury, Venus, Earth, and Mars—are composed largely of rock and are known as the terrestrial (Earth-like) planets. As Figure 1.11 suggests, we can apply what we know about Earth to the study of the other terrestrial planets because they have similar compositions. Similarly, we can apply what we learn about Mercury, Venus, and Mars back to our study of Earth. The outer four planets—Jupiter, Saturn, Uranus, and Neptune—are much more massive and are composed largely or entirely of gas, and are known as the gas giants.

Figure 1.11. The Curiosity rover on Mars, one of the four terrestrial planets. Many of the geologic features that occur on Earth, such as volcanoes, sand dunes, and stream channels, can also be studied on Mars.

1.2.2 Earth—A "Just Right" Planet

Most places in the solar system, or in the universe as a whole, are quite hostile to complex life. Complex life is life that is more sophisticated than bacteria, and includes all plants and animals. In most places in the universe, the temperatures are too hot or too cold, or there isn't water, or the right elements—such as carbon—aren't present, or there is too much damaging electromagnetic radiation for living organisms to thrive.

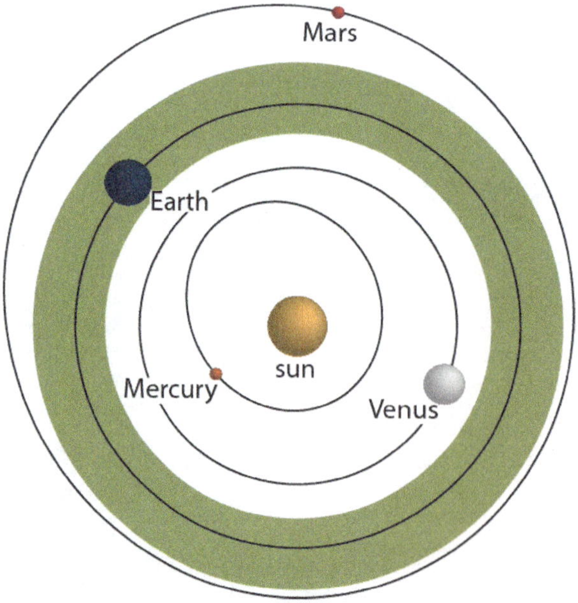

Figure 1.12. Our solar system's habitable zone, tinted in green. There is debate about where exactly the inner and outer limits of the habitable zone are. The sun and planets are not drawn to scale.

Within the solar system, Earth occupies a special location. If Earth were somewhat closer to the sun, the temperature would be hot enough to boil the oceans. Living organisms, from bacteria to humans, are dependent on liquid water to exist. Even on Venus, the next planet closer to the sun, the temperature on the surface is around 460°C (860°F)! On the other hand, if Earth were farther away from the sun, it would be cold enough to freeze all water on the surface. As an example, consider Mars, with an average surface temperature of around –55°C (–67°F), though temperatures can rise above freezing (0°C) on a summer day near the equator. As far as we know, Mars is currently a lifeless planet. It certainly cannot support the abundance of life that exists on Earth.

The region around a star in which liquid water can exist on the surface of planets—and which therefore can support advanced life—is known as the *habitable zone*, illustrated in Figure 1.12. Determining the inner and outer limits of the habitable zone is not a straightforward task, but certainly Venus is too close to the sun, and Mars is at or near the outer edge of our solar system's habitable zone. Sometimes astronomers refer to planets in the habitable zone as "Goldilocks planets," in which the temperature is neither too hot, nor too cold, but just right. There may be places outside of the habitable zone where conditions are just right for life—such as in warm water beneath the surface of some of the moons of Jupiter—but it is considered unlikely by most scientists that these environments could support life more sophisticated than microscopic bacteria.

Earth is also "just right" in other ways:

1. Earth seems to be an ideal size. If it were considerably smaller, it would not have strong enough gravity to hold onto most of its atmosphere. Mars has only 10.7 percent of the mass of Earth, and has a very thin atmosphere. If, on the other hand, Earth were considerably larger, it would probably have a much denser atmosphere due to stronger gravity. This would likely lead to much higher surface temperatures.
2. Earth has a good amount of water. There is enough water to support life, but not so much to make Earth completely covered with water.
3. Earth seems to have just the right chemical composition. For example, Earth has a small amount of carbon, which is an essential element for all of the primary molecules of life, such as proteins, sugars, and DNA. But if it had considerably more carbon, the composition of both the solid Earth and the atmosphere would be radically different, and inhospitable to life. Earth also has an iron core, which causes Earth's magnetic field and helps to protect the surface from harmful radiation from space (the solar wind). The chemical composition of Earth's crust also seems to be just right for the long-term maintenance of life.
4. Many scientists believe that gravitational interactions between Earth and the moon—the same interactions that cause ocean tides—keep Earth's axis tilted at a fairly constant angle near 23.5°. If Earth didn't have such a large moon—all other moons in the solar system are small compared to the size of their parent planets—occasional changes in the angle of Earth's tilt could cause catastrophic changes to the climate, which would make the continued existence of complex life difficult. We will take a closer look at Earth's axis in the next section.
5. Earth has plate tectonics, which, as you will learn in Chapter 6, is the process that moves continents and other parts of the lithosphere around on Earth's surface.

It turns out that plate tectonics is a process which helps to make Earth a suitable home for the flourishing of life. Geologists believe that a number of factors have to be just right for plate tectonics to occur, such as the presence of oceans, and having the right amount of radioactivity in the rocks of the lithosphere.

There are many more ways in which the universe as a whole, and our planet in particular, seem to be purposefully designed and fine-tuned for life. Psalm 19:1 states that "The heavens declare the glory of God, and the sky above proclaims his handiwork." To Christian Earth scientists, it is not only the heavens and sky that declare God's glory, but all of his creation, including everything in the lithosphere, hydrosphere, atmosphere, and biosphere. All Christians believe the entire universe testifies to its loving and providential Creator, our wise God who designed the universe to be inhabited by a spectacular variety of complex life, including us.

1.2.3 Earth in the Galaxy and Universe

On a very dark night, one can see up to two or three thousand stars, which is only a very tiny fraction of the total number of stars in the galaxy. A galaxy is a massive system of stars, gravitationally bound to one another. Our sun is but one of perhaps 200 billion (200,000,000,000) stars in the Milky Way galaxy, the spiral galaxy in which our solar system is located. If we were able to travel outside of the Milky Way galaxy—which would take millions of years to do with our current spacecraft—it would look something like the galaxy shown in Figure 1.13. As Figure 1.14 illustrates, our solar system is located at the edge of a spiral arm, roughly half way from the center of the galaxy to the outer edge. Our sun is orbiting the center of the galaxy with a velocity of about 800,000 kilometers per hour, but even moving that fast, it takes over 200,000,000 years for the sun to make one revolution around the galactic core.

Just as there is a habitable zone in the solar system, where conditions are right for life to flourish, so there seems to be a galactic habitable zone in spiral galaxies where conditions are right for life to exist. Near the galactic core, stars are closer to each other than they are in our neighborhood of the galaxy, and it is believed that catastrophic events, such as stars passing too close to one another and disrupting planetary orbits due to gravitational attraction, would make the entire central region of the galaxy inhospitable for complex life. On the other hand, it seems that solar systems near the edges of spiral galaxies would not contain a high enough proportion of atoms of elements heavier than hydrogen and helium

Figure 1.13. Our Milky Way galaxy is a spiral galaxy, like the Andromeda galaxy pictured here.

to have terrestrial planets, which are made of heavier elements such as iron, silicon, and oxygen.

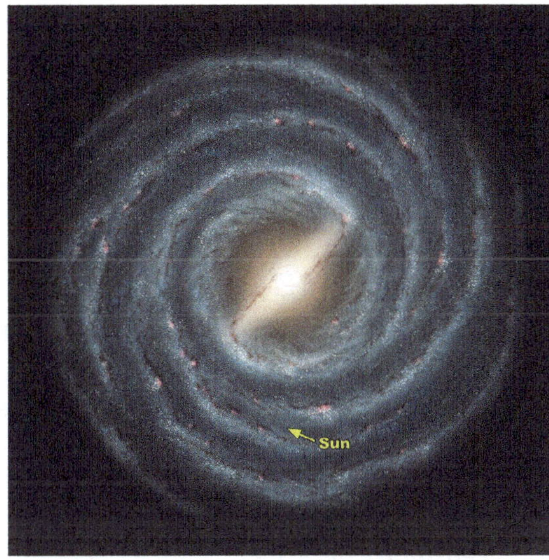

Figure 1.14. An artist's conception of the Milky Way galaxy, showing the location of our sun and solar system.

The Milky Way galaxy is just part of the much larger *universe.* The universe is made up of all matter and energy that exists, as well as space and time themselves. This includes everything we can see with our most powerful telescopes, such as the distant galaxies shown in Figure 1.15. The Milky Way galaxy is just one of hundreds of billions of galaxies. If there are 200,000,000,000 stars in our galaxy, and at least 100,000,000,000 observable galaxies in the universe, then how many stars are there in the entire universe? At a minimum, that number would be 20,000,000,000,000,000,000,000 stars.

This enormous number is virtually incomprehensible to us. It would take over 600 trillion years to count 20,000,000,000,000,000,000,000 stars, taking one second

Figure 1.15. The universe is made up of many billions of galaxies. Almost every object in this image, taken by the Hubble Space Telescope, is a galaxy composed of billions of stars.

per star. Because the universe is so enormous, and our sun and planet are so small in comparison, it would be easy to conclude that we humans are rather insignificant. King David, as recorded in Psalm 8:3–4, also looked at the creation and asked,

> *When I look at your heavens, the work of your fingers,*
> *the moon and the stars, which you have set in place,*
> *what is man that you are mindful of him,*
> *and the son of man that you care for him?*

We are, indeed, very small compared to the rest of the creation. But God intended it this way; the vastness of the universe speaks of the bigness of God even more than it speaks of the smallness of the human race. I just stated that it would take many trillions of years for us to count all the stars in the universe, but according to Isaiah 40:26, God has a name for each of those stars. Things that are far beyond our comprehension and capabilities are easy for our powerful Creator!

Learning Check 1.2

1. Contrast the geocentric model of the solar system with the heliocentric model.
2. Describe Earth's location in space, relative to the sun and solar system, the galaxy, and the universe.
3. In what ways does Earth seem to be "just right" for complex life?
4. What is meant by a habitable zone, for both the solar system and galaxy as a whole?
5. In what ways does Earth seem to be a very small place in the universe? What does this tell us about ourselves? What does it tell us about God?

1.3 Earth's Orbit and the Seasons

1.3.1 Earth's Orbit

Earth takes 365.24 days to make one complete orbit around the sun. This orbit, however, is not circular. Instead, Earth's orbit around the sun is elliptical, which means that Earth is slightly closer to the sun at some times, and farther away at others. At its closest, Earth is about 147 million kilometers from the sun, and at its farthest it is about 152 million kilometers, for an average of about 150 million kilometers (93,000,000 miles). The point on the orbit of a planet, asteroid, or comet around the sun where it is closest to the sun is called the *perihelion*, and the point where it is farthest from the sun is called the *aphelion*, as shown in Figure 1.16. The differ-

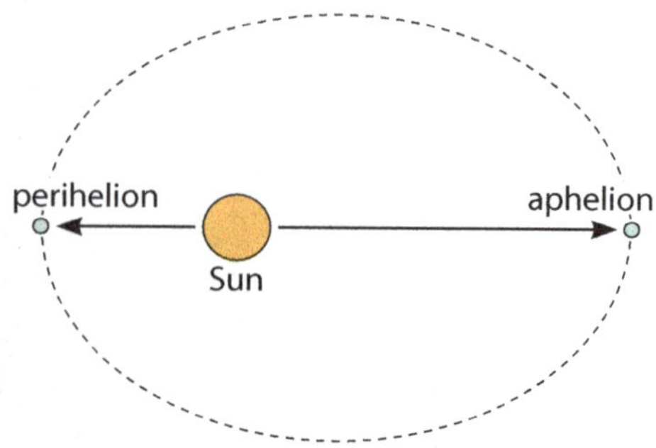

Figure 1.16. For an elliptical orbit, the aphelion is where a planet is farthest from the sun, and the perihelion is where a body is closest to the sun.

ence between the perihelion and aphelion for Earth's orbit is not all that great, and from a distance Earth's orbit would appear to be nearly circular. One interesting thing about this elliptical orbit is that Earth is closest to the sun (perihelion) during the first week of January, and farthest from the sun (aphelion) the first week of July. This means Earth is actually slightly closer to the sun when it is winter in the Northern Hemisphere and slightly farther away from the sun in the summer. What this tells us is that the distance Earth is from the sun is not the primary cause of seasons on Earth. This distance does have a minor effect on seasons—the Northern Hemisphere winter is slightly warmer than it would be if Earth's orbit were circular, and Northern Hemisphere summer is slightly cooler than it would otherwise be—but the main cause of Earth's seasons is the tilt of Earth's axis.

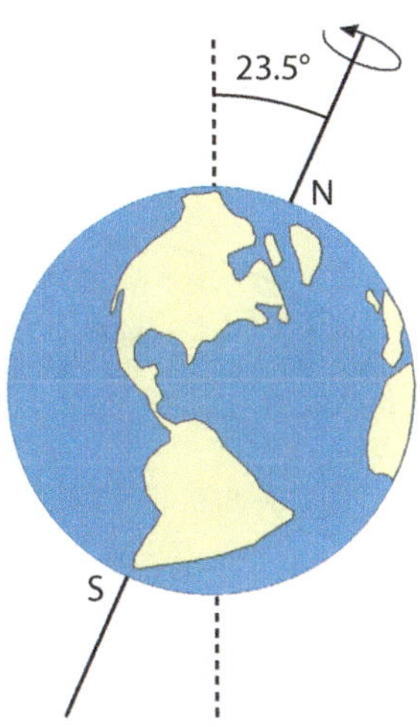

Figure 1.17. Earth's axis is tilted 23.5° from vertical, relative to the ecliptic.

1.3.2 Solstices and Equinoxes

The plane in which Earth orbits the sun is called the *ecliptic*. As Earth orbits the sun, its axis of rotation is not perpendicular to the ecliptic, but is tilted at an angle of about 23.5° as Figure 1.17 shows. As Earth orbits the sun, its axis always points in the same direction. In Figure 1.13, you can see that no matter where Earth is in its orbit, its axis is inclined 23.5° to the right. As Earth completes its annual trip around the sun, different parts of the planet receive different amounts of sunlight. At the far left of Figure 1.18, on about June 21, the Northern Hemisphere is tilted towards the sun and receives more direct sunlight than does the Southern Hemisphere. The *summer solstice* is the moment of time when the sun is highest in the sky in the hemisphere, and is considered to be the first day of summer. In general, the summer solstice is the day of the year with the most hours of sunlight. The *winter solstice* occurs on about December 21 in the Northern Hemisphere. This is when the sun is lowest in the sky in the hemisphere, and is the day with the least hours of sunlight. At the time the summer solstice occurs in the Northern Hemisphere, the winter solstice occurs in the Southern Hemisphere. About halfway between the solstices are the equinoxes, when Earth's axis neither tilts away nor toward the sun, and

Figure 1.18. As Earth orbits the sun, its axis is always pointed the same direction. This means that for part of the year, the North Pole is pointing more towards the sun, and at the opposite part of the year, the North Pole is pointing away from the sun.

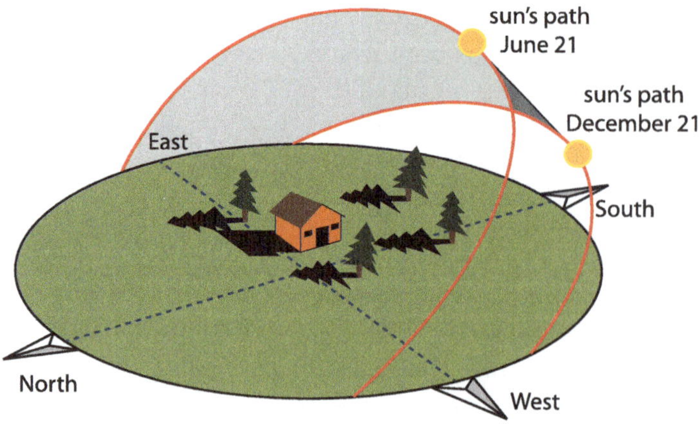

Figure 1.19. The path the sun takes across the sky in the Northern Hemisphere at the latitude of the central United States on the summer solstice (June 21) and winter solstice (December 21).

there is approximately an equal amount of daylight and night time. The *vernal equinox* is the equinox that occurs between the winter solstice and summer solstice, on about March 22 in the Northern Hemisphere. The *autumnal equinox* occurs between the summer solstice and winter solstice, on about September 22 in the Northern Hemisphere.

The temperature on Earth's surface depends on the amount of heat received from the sun, which is received in the form of electromagnetic radiation. Earth is always gaining energy from the sun, but Earth also loses heat energy into space at the same time. If Earth did not lose heat to balance out the energy it receives from the sun, it would get hotter and hotter and eventually melt! The electromagnetic radiation from the sun is primarily visible light, which is absorbed by Earth's surface and atmosphere, leading to warming. The warm Earth, at the same time, re-radiates electromagnetic radiation back out into space, mainly as infrared radiation. We can see this balance swinging back and forth on a daily basis, as the sun warms Earth's surface during the day, but then the temperature drops at night. We also see this happening on a longer time scale, as the Northern Hemisphere warms up through the spring and summer, and cools off through fall and winter. It is all about which is greater at the time: energy received from the sun, or energy radiated back into space from Earth.

Figure 1.20. The flashlights illustrate the difference between having the sun high in the sky (summer) and low in the sky (winter). In the summer, energy is concentrated in a smaller area, but in winter, energy from the sun is spread out over a larger area, resulting in less heating of the ground.

There are two primary reasons why it is hotter in the summer than in the winter. Both reasons are related to the path the sun takes across the sky. We think of the sun as rising in the east and setting in the west. But as Figure 1.19 shows, in the summer the sun actually rises in the northeast and takes a long, high path across the sky before setting in the northwest (from a Northern Hemisphere perspective). In the winter, on the other hand, the sun rises in the southeast, and takes a much shorter, lower path across the sky before setting in the southwest.

From these observations, it becomes apparent why it is hot in the summer. First of all, the sun is in the sky for a long time on a summer day, which means there is a long period of time for Earth's surface to absorb rays from the sun. Second, the sun is higher in the sky on a summer day, which means that sunlight is concentrated on a smaller area than in the winter, illustrated with flashlight beams in Figure 1.20. The opposite is true in winter; the sun is in the sky for a shorter period of time, which means there is less time for Earth's surface to absorb sunlight, and sunlight hits Earth's surface at a lower angle, which spreads out the energy from the sun over a greater area.

1.3.3 The Tropics and Polar Regions

Near Earth's equator, changes in the sun's position in the sky don't have as great of an effect on temperatures. Whether it is summer solstice, winter solstice, or an equinox, the sun climbs up high in the sky every day, and this part of Earth is generally warm year-round. Between latitudes 23.5°N and 23.5°S, there is at least one day in the year when the sun is directly overhead at noon.[1] At the latitude 23.5°N, the sun climbs up to directly overhead at noon on the Northern Hemisphere summer solstice. As shown in Figure 1.21, this latitude is known as the *Tropic of Cancer*. This latitude is the farthest north that the sun can be seen directly overhead. The southern equivalent, at 23.5°S, is the *Tropic of Capricorn*. The region between the Tropic of Cancer and the Tropic of Capricorn is known as the *tropics*, and except in mountainous areas, is warm all year. In the United States, only Hawaii is in the top-

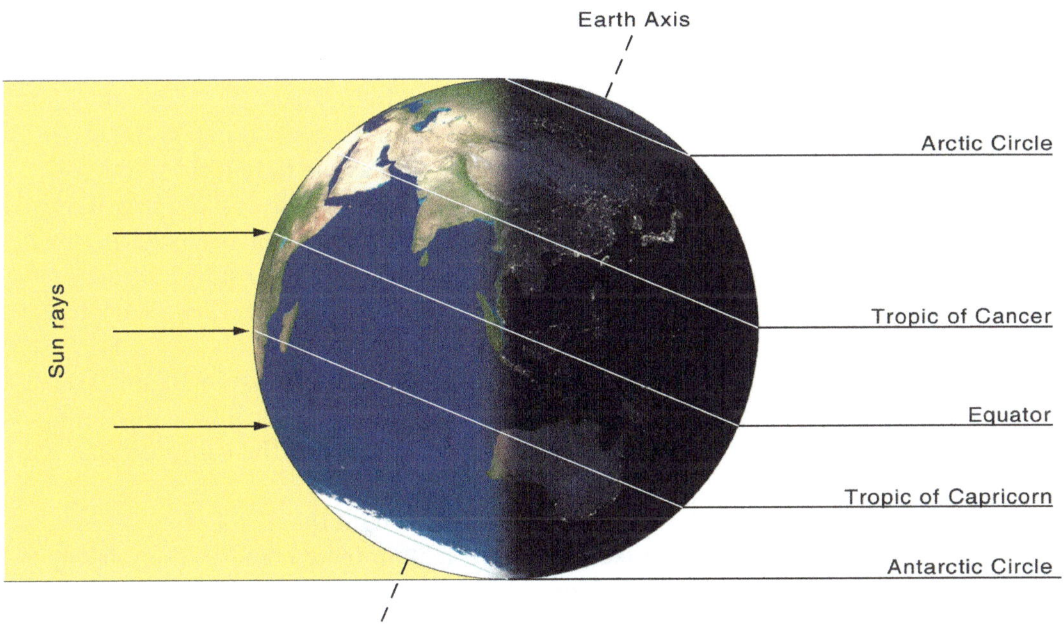

Figure 1.21. Earth at Northern Hemisphere winter solstice, showing the Tropic of Capricorn, Tropic of Cancer, Arctic Circle, and Antarctic Circle.

1 Latitude and longitude will be discussed in Chapter 2.

ics. In many tropical areas, the main difference in seasons is not between hot and cold, but between a wet season and a dry season.

The situation is quite different in polar regions. On the day of the Northern Hemisphere summer solstice, the sun does not set at any location farther north than the *Arctic Circle*, which is at about 66.5°N. These regions are sometimes referred to as "the land of the midnight sun." But the opposite situation occurs at the winter solstice, when the sun does not rise above the horizon north of the Arctic Circle. The Southern Hemisphere equivalent of the Arctic Circle is the *Antarctic Circle*, which is at approximately 66.5°S. Despite having up to 24 hours of sunlight, polar regions do not get hot in the summer because the sun is always low in the sky.

Learning Check 1.3

1. Calculate the approximate speed at which Earth moves in its orbit around the sun. You can simplify the problem by assuming that Earth's orbit around the sun is circular.
2. What effect does the elliptical orbit of the sun have on Earth's seasons?
3. Explain how the tilt of Earth's axis causes seasons.
4. Speculate as to what seasons would be like if Earth's axis were not tilted, and what seasons would be like if Earth's axis were tilted at 45° instead of 23.5°.

1.4 Phases of the Moon

There are nights when the moon shines bright in the sky; bright enough to walk outside without the need for artificial lighting. There are other nights when it is pitch black and there is no moon above, but thousands of stars shine in the dark sky. As the moon orbits Earth, its appearance, as well as the time it rises and sets, changes from day to day. The *phase* of the moon, or *lunar phase*, is the shape of the sunlit portion of the moon's face as seen from Earth. As the moon orbits Earth, the same hemisphere of the moon always faces us, but part of the near side of the moon is in sunlight, and part of it is in darkness. The phases of the 29.5-day lunar cycle are outlined in Table 1.1.

A good way of visualizing how lunar phases work is shown in Figure 1.22, which shows the moon revolving around Earth in a counter-clockwise direction. The phases of the moon are determined by the positions of the sun, Earth, and moon.

The moon emits no light of its own; we only see light that is reflected off of it. Note that the half of the moon facing the sun is illuminated, and the side facing away from the sun is dark. As the moon orbits Earth, varying amounts of light and dark are visible from Earth's surface.

When the moon is in the position represented by the right side of Figure 1.17, the side of the moon facing Earth is completely in darkness, while the side of the moon facing away from Earth is fully illuminated. This is called the *new moon*.

Phase	Moonrise	Moonset	Viewing
 New Moon	Sunrise	Sunset	Moon not visible
 Waxing Crescent	Just after sunrise	Just after sunset	Moon visible in the west just after sunset
 First Quarter	Noon	Midnight	Moon visible in afternoon and early evening
 Waxing Gibbous	Late afternoon	After midnight	Moon visible most of the night
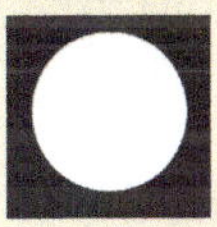 Full Moon	Sunset	Sunrise	Moon visible all night
 Waning Gibbous	Before midnight	Mid-morning	Moon visible late night until mid-morning
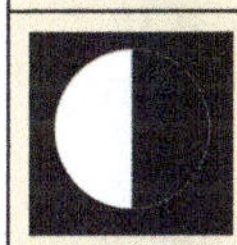 Last Quarter	Midnight	Noon	Moon visible after midnight until noon
 Waning Crescent	Just before sunrise	Just before sunset	Moon visible in the east just before sunrise

Table 1.1. Phases of the moon, with moonrise and moonset times.

When the moon is new, it is close in the sky to the sun, so it rises in the morning, sets in the evening, and is invisible to us. A couple of days after the new moon, the moon has advanced in its orbit around Earth, and a thin sliver is visible in the western sky just after sunset. This is called the *waxing crescent*; the term waxing means increasing or growing. If you observe the moon from night to night, you will see that the line that separates light from dark slowly moves across the moon's face from right to left. About a week after the new moon, the moon has completed a quarter of its orbit around Earth, and it appears half illuminated. Rather than calling this a half moon, however, it is referred to as a quarter moon—in this case, the *first quarter* (top of Figure 1.22). After the first quarter, the near side becomes more fully illuminated, with more than half but less than all of the near side showing, a phase called the *waxing gibbous*. A full two weeks after the new moon, the moon is fully

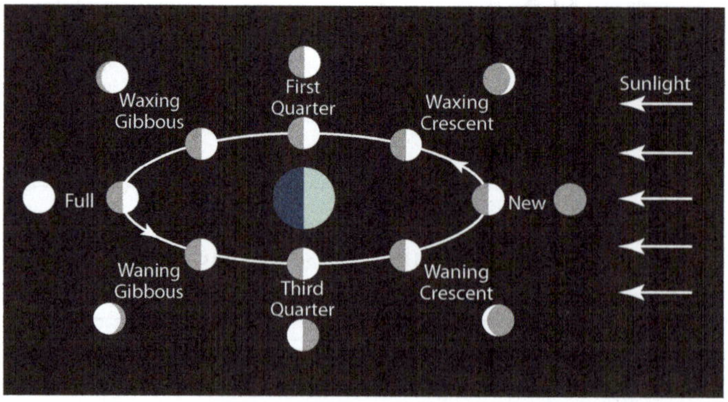

Figure 1.22. As the moon orbits Earth, varying amounts of the near side of the moon become illuminated by the sun, causing lunar phases.

illuminated when viewed from Earth, and we see a *full moon* (far left of Figure 1.22).

During the following two weeks, the process reverses (lower part of Figure 1.22). A few days after the full moon, darkness starts to slowly creep across the moon's face from right to left, leading to the waning gibbous phase. Waning means decreasing, the opposite of waxing. The *last quarter* occurs a week after the full moon, and a few days later, the *waning crescent* appears in the sky shortly before sunrise. After 29.5 days, the cycle repeats itself, starting with a new moon.

Something many people don't realize is that the moon is in the sky during the day time just as often it is in the sky during the night time. We don't notice the daytime moon as often for a couple reasons. First of all, the sky is bright during the day, and it is easy to miss a pale crescent or quarter moon, or even a gibbous moon, against the blue sky. Second, even though the moon is in the sky when it is at or near the new moon phase, there is simply not enough of the sunlit side of the moon visible for us to see.

The time of day that the moon rises and sets goes hand in hand with the phases of the moon. For example, the full moon is on the opposite side of Earth from where the sun is, so the full moon rises roughly when the sun sets and sets roughly when the sun rises. Each day, the moon rises about 50 minutes later; thus three days after the full moon, the waning gibbous moon rises about 2.5 hours (150 minutes) later. Table 1.1 summarizes when the moon rises and sets for different phases.

Learning Check 1.4

1. Explain why the moon has phases.
2. Why does the full moon rise at roughly the same time that the sun sets?
3. If you were on the moon, how would the appearance of Earth change throughout the 29.5-day lunar cycle?

1.5 Eclipses

In addition to phases, the orbit of the moon around Earth causes another interesting phenomenon: eclipses. An eclipse occurs when one celestial body, such as

Earth or the moon, blocks the sun from another celestial body.

1.5.1 Solar Eclipses

Figure 1.23. Partial solar eclipse. Looking directly at a partial eclipse can damage your eyes!

A *solar eclipse* occurs when the moon passes in front of the sun, causing the moon's shadow to fall on Earth. In a *partial solar eclipse*, the moon only covers part of the sun's disk, as shown in Figure 1.23. In the case of a *total solar eclipse*, the moon completely covers the sun, as in Figure 1.24. During a total solar eclipse, the sky becomes dark, but the total part of the eclipse lasts for only a few minutes. Solar eclipses can only happen when the moon is between Earth and the sun, when there is a new moon, illustrated in Figure 1.25.

A shadow from a light source, such as the sun or a light bulb, will have multiple parts to it. The darkest, inner part of the shadow is called the *umbra*. During a solar eclipse, the moon completely blocks the sun within the umbra. The *penumbra* is the part of the shadow that is not completely dark. Within the penumbra, the sun is completely blocked, so a viewer sees a partial solar eclipse. The umbra and penumbra are easy to demonstrate, using a table lamp and a disk of some type, as shown in Figure 1.26.

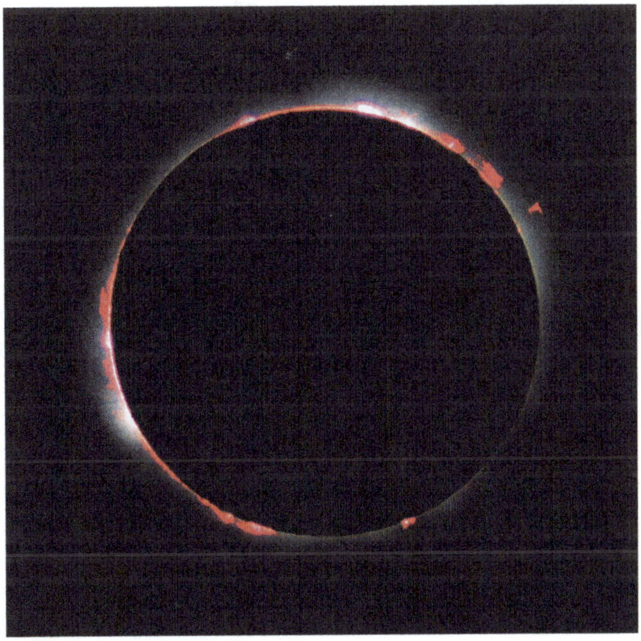

Figure 1.24. Total solar eclipse. This is the only time that the solar atmosphere is visible from Earth.

Total solar eclipses are fairly rare events, and most people have never seen one. (I have seen a total solar eclipse once, in 1979). Looking back at Figure 1.22 (lunar phases), one might think that a solar eclipse should occur every month. But the moon orbits Earth in a plane that is about 5° tilted relative to the ecliptic, the plane in which Earth orbits the sun. This means that in most months, the new moon passes either above or below the sun in the sky rather than right in front of it. In some years, there are no total solar eclipses. In other years, there may be up to five solar eclipses, although this only

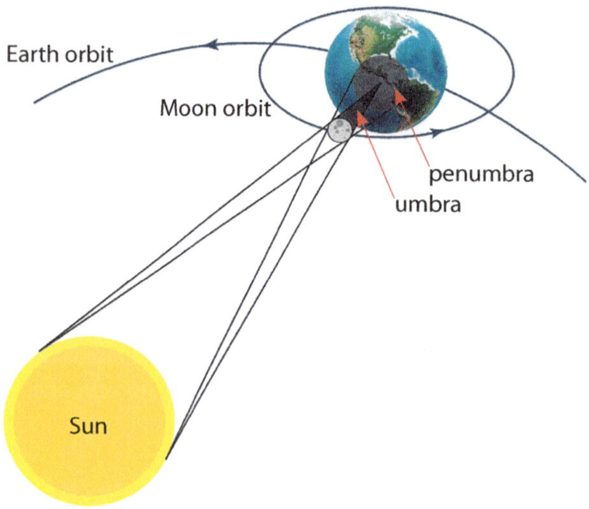

Figure 1.25. A solar eclipse occurs when the moon blocks the light of the sun. Viewers located in the umbra, which is very small on Earth, see a total solar eclipse, and viewers located in the penumbra see a partial solar eclipse.

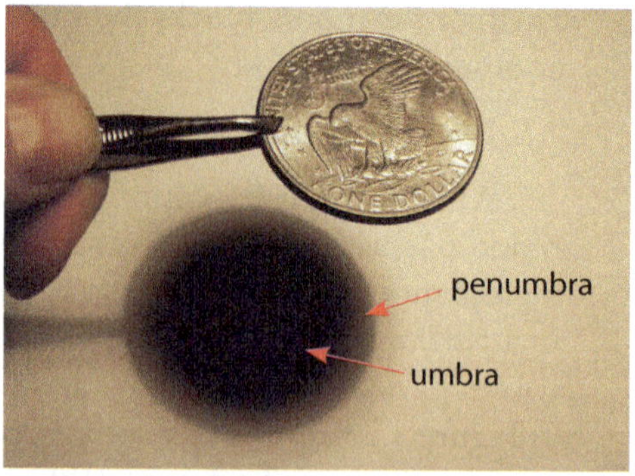

Figure 1.26. Umbra and penumbra. An ant within the umbra would not be able to see the light bulb, and an ant in the penumbra would be able to see part of the light bulb.

happens rarely. Because of the small size of the umbra on Earth, total solar eclipses are only visible in narrow bands on Earth's surface.

There is actually more to the sun than just the bright disk we normally see in the sky. When that disk is blocked, we can see the solar atmosphere, which is seen as a thin ring of pink light around the eclipse plus a faint white glow that extends further out (Figure 1.24). The solar atmosphere is made of hot glowing plasma, which is an ionized gas.

The sun is much larger than the moon, but it is also much farther away. The result is that the moon and the sun appear to be almost exactly the same size in the sky. It is truly amazing that we live on a planet that has total solar eclipses in which the moon almost perfectly covers the sun. This involves having a moon that is just the right size at just the right distance from Earth. Being able to see solar eclipses is not just something that is cool, but something that has been critical in the development of science. For example, total solar eclipses helped us to understand the composition of the sun, and helped to confirm the validity of Albert Einstein's theory of general relativity. One could argue that there is nothing very spe-

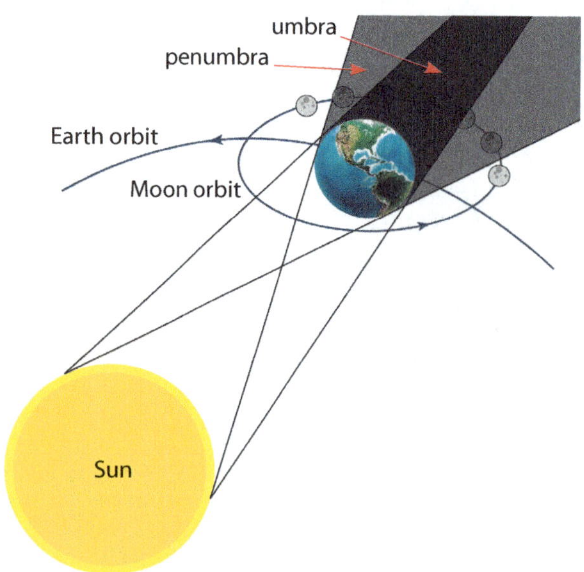

Figure 1.27. A total lunar eclipse occurs when the moon passes through Earth's umbra.

cial about our eclipses, and that there are natural explanations for the size and location of the moon. But as a Christian, I look at this as one more example of God's gracious providence that allows us to understand his creation more completely.

1.5.2 Lunar Eclipses

Figure 1.28. The moon during a total lunar eclipse. The reddish glow is caused by sunlight bending through Earth's atmosphere.

Lunar eclipses occur when Earth's shadow covers the face of the moon. This can happen when the moon is full, when Earth is between the moon and the sun, but does not happen every time the moon is full. Again, because the moon's orbit around Earth is inclined at 5° to the ecliptic, the moon usually misses Earth's shadow as it orbits.

Like all shadows, Earth's shadow has an umbra and penumbra, illustrated in Figure 1.27. When the moon passes through Earth's penumbra, it becomes slightly darker, but this dimming can be rather subtle. A *total lunar eclipse* occurs when the moon passes completely into Earth's umbra. During a total lunar eclipse, the moon does not become completely dark, even though Earth completely blocks the sun. If you were standing on the moon at this time, you would see Earth completely blocking the sun, with a brilliant red sunset making a ring around the entire Earth. This red light actually illuminates the moon during a total lunar eclipse, giving the moon the reddish glow shown in Figure 1.28. For this reason, the eclipsed moon is often referred to as a "blood moon."

Earth's shadow is much larger than the moon's shadow. Because of this, Earth's umbra can completely cover the moon (unlike the moon's umbra, which darkens only a small area on Earth). During a lunar eclipse, anyone on the night side of Earth can see the full moon, and therefore can see the eclipse occurring. People on the day side of Earth miss the eclipse.

Learning Check 1.5

1. Describe how the sun, Earth, and moon must be arranged for solar and lunar eclipses to occur.
2. During a solar eclipse, why do people in some places see a partial eclipse, and others see a total eclipse?
3. Explain why there aren't a solar eclipse and a lunar eclipse every month.

1.6 Calendars

Closely related to Earth's revolution around the sun and the moon's orbit around Earth is the topic of calendars. The calendar we use in much of the world today, including in all of Western society, is based on Earth's orbit around the sun. Before examining our Western calendar, however, we will take a brief look at some other calendars that have been used in other parts of the world.

1.6.1 The Hebrew Calendar

Most calendars are systems that divide the passage of time into years, months, and days, but the details vary. We will start by looking at the Hebrew calendar, used by the Jews in Biblical times as well as in modern Israel. The Jewish concept of "day" is reflected in Genesis 1, where the phrase "And there was evening and there was morning, the first day" (Genesis 1:5) is repeated for each of the six days of creation. As reflected in the wording of this phrase in Genesis 1, the Hebrew day begins at sunset rather than at midnight.

The first chapter of Genesis also lays a foundation for other aspects of the Hebrew calendar, which is the tracking of seasons and years. Genesis 1:14-15 reads:

> *And God said, 'Let there be lights in the expanse of the heavens to separate the day from the night. And let them be for signs and for seasons, and for days and years, and let them be lights in the expanse of the heavens to give light upon Earth.' And it was so. And God made the two great lights—the greater light to rule the day and the lesser light to rule the night—and the stars.*

The Hebrew calendar is an example of a lunisolar calendar, which is based on both the phases of the moon (the "lesser light") and the orbit of Earth around the sun (the "greater light"). The moon orbits Earth once every 29.5 days, so each month in the Hebrew calendar has either 29 or 30 days, giving a total of 354 days in twelve months. Each month starts with the new moon. Earth does not completely orbit the sun in 354 days, and the Hebrew calendar compensates for this by adding a thirteenth month in seven out of every nineteen years. This adjustment is necessary so the months occur in a consistent time of the year. Without this correction the seasons—and associated religious holidays—would occur about eleven days earlier each year. For example, the original Passover, as recorded in Exodus, occurred in the spring. Without the addition of a leap month every few years, Passover would occur in April for about three years, then in March for a few years, and so on. The calendar was also used for agricultural purposes as well, such as planting and harvesting.

The numbering of years for the Hebrew calendar is based on a medieval calculation for the creation of the world. The Jewish New Year celebration, Rosh Hashanah, usually occurs in September, with the Hebrew year 5776 beginning in September of 2015.

1.6.2 The Islamic Calendar

The Islamic calendar is used by Muslims throughout the world for tracking religious holidays, and as the day-to-day calendar for business and government in many Islamic countries. The Islamic calendar is a lunar calendar in which months are based strictly on the phases of the moon. Each new day begins at sunset. Like the Hebrew calendar, each month in the Islamic calendar has either 29 or 30 days. A new month begins at the first sighting of the thin crescent moon the day after the new moon, shown in Figure 1.29. This determination is usually made by religious authorities for a given country, and if they do not see the crescent moon for some reason (such as cloud cover), then the new month doesn't begin until the next evening.

Traditionally, it has been difficult to create a calendar for future months of the Islamic year, or to make a firm appointment for a given date in a future month. A person could not predict in advance whether the current month would have 29 or 30 days; it depended on whether the crescent moon was visible in the western sky at the end of the 29th day. Some Islamic countries now allow astronomical calculations to determine when new months will begin, but others continue to rely on actual observations of the sky.

The Islamic calendar has twelve months that average 29.5 days long, for a typical total of 354 days in a year. Unlike the Hebrew calendar and other lunisolar calendars, there are no leap months. In North America and Europe, we associate certain months and holidays with certain seasons—Christmas is in winter, and July is a summer month. Islamic months and holidays, on the other hand, come eleven days earlier each year, so the months and holidays migrate through the seasons. For example, the month of Ramadan—the month in which adult Muslims are obligated to fast from food and water from sunrise to sunset every day—occurs in the summer for a few years, then in the spring for a few years, and so on. The Islamic year 1438 begins in October of 2016.

Figure 1.29. Months in the Islamic calendar begin with the first sighting of the waxing crescent moon.

1.6.3 Western Calendars

There have been two main calendars used in Western society over the past 2,000 years, and they are both solar calendars, based primarily on Earth's orbit around the sun. The first of these is the Julian calendar, named after Julius Caesar, who introduced it in 46 BC. It was very similar to the calendar we use today. There were twelve months with either 30 or 31 days, except February which normally had 28 days, making a total of 365 days in a year. Every fourth year was declared to be a leap year, with a 29th day in February.

Originally, the Julian calendar had no consistent way of numbering years. Sometimes a year would be indicated by the reign of an emperor, as in Luke 3:1, which begins with "In the fifteenth year of the reign of Tiberius Caesar." The idea of numbering the years starting with the birth of Jesus Christ—Anno Domini (year of the Lord), or AD—originated in about 525 AD, and gradually spread throughout the Christian world. Unfortunately, the calculations for determining the birth of Christ were off by a few years—Jesus was probably born around 4–6 BC (BC means Before Christ). There was no year zero; 1 BC was followed immediately by 1 AD.

There was a minor problem with the Julian calendar that grew with time. The average length of a year in the Julian calendar was 365.25 days, but Earth actually takes about 365.24 days to orbit the sun. Over a period of many centuries, this meant that months and religious holidays began to migrate relative to the seasons. Given enough time, the observance of Easter eventually would have become a winter, rather than a spring, event. In 1582, Pope Gregory XIII introduced a revision to the Julian calendar, which is known as the Gregorian calendar. This is the calendar we use today, and has become the international standard, even in many countries that use other calendars for religious or cultural purposes.

The main difference between the Julian and Gregorian calendars is a reduction in the number of leap years in the Gregorian calendar. With the Julian calendar, any year divisible by four was a leap year. The Gregorian calendar requires that years divisible by 100 must also be divisible by 400 in order to be a leap year. The years 1700, 1800, and 1900 would have been leap years using the Julian calendar, but not leap years using the Gregorian calendar. The year 2000 was a leap year because it is divisible by 400.

Some Christian holidays have fixed dates, such as Christmas, which falls on 25 December on the Gregorian calendar. Other church holidays, such as Easter, are actually based on a lunisolar calendar. Easter occurs on the first Sunday after the first full moon following the March equinox—the vernal equinox in the Northern Hemisphere.

Learning Check 1.6

1. Contrast the Hebrew and Islamic calendars, distinguishing between the lunar calendar and the lunisolar calendar.
2. Describe the features of the Julian and Gregorian calendars. Explain why it was necessary to switch from one to the other.

Chapter 1 Exercises

Answer each of the questions below as completely as you can. Write your responses in complete sentences unless instructed otherwise.

1. Give a definition for each of the four major Earth systems and describe how these systems interact with each other.
2. Draw a sketch of our solar system, showing the sun, planets, and habitable zone.
3. What are some ways in which Earth appears to be specially designed for living organisms to thrive?
4. Describe what is meant by the term "galactic habitable zone."
5. Describe Earth's orbit around the sun and explain why the shape of this orbit has little effect on Earth's seasons.
6. Draw a sketch like Figure 1.13 and use it to explain what causes Earth's seasons.
7. Explain why, in general, summer days are hot and winter days are cold.
8. Predict what Earth's days and seasons would be like if Earth's axis were tilted at close to 90° rather than 23.5°.
9. Draw a sketch of Earth, showing the location of the Tropics of Cancer and Capricorn, and Arctic and Antarctic Circles.
10. Describe the path the sun would take across the sky viewed from the North Pole on the summer solstice, autumnal equinox, and winter solstice.
11. Draw sketches of the eight phases of the moon, in order, from the new moon through the waning crescent.
12. Draw a sketch like Figure 1.17 and use it to explain the cause of lunar phases.
13. When people on Earth see a quarter moon, what would Earth look like from the moon?
14. Draw a sketch of a total solar eclipse with labels for the sun, moon, Earth, umbra, and penumbra.
15. Explain why solar eclipses are only visible in limited geographic areas, but lunar eclipses are visible from the entire night side of Earth.
16. Describe the similarities between the Hebrew and Islamic calendars, and then explain what the biggest difference is between how they work.
17. Compare the Julian and Gregorian calendars, and explain how the Gregorian calendar corrected a problem with the Julian calendar.

Chapter 2

Visualizing Earth

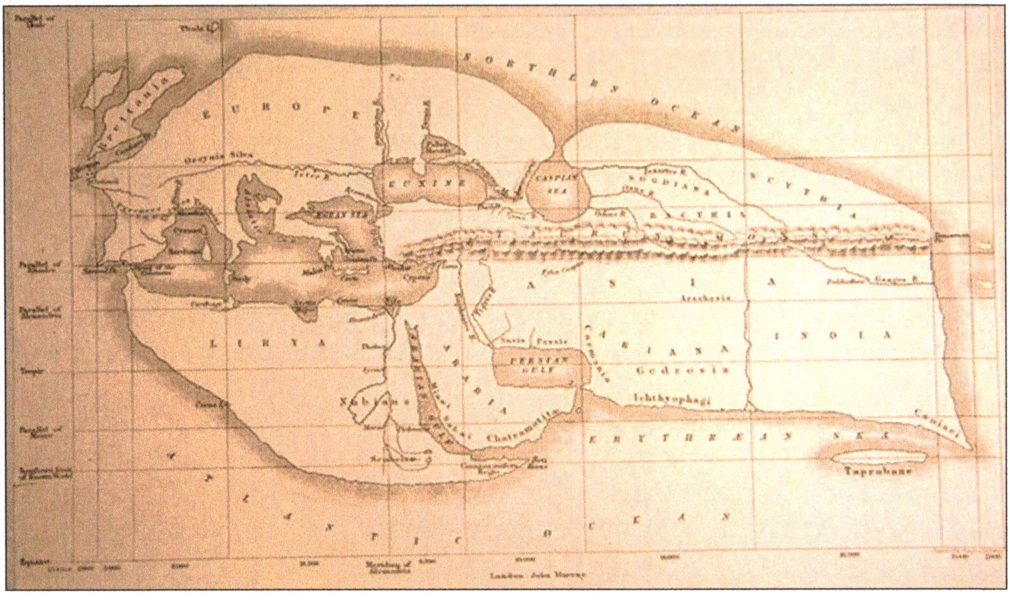

The ancient Greek scholar Eratosthenes of Cyrene, who lived from about 276 to 195 B.C., is considered by many to be the first true geographer. Other Greek philosophers, such as Aristotle, had demonstrated that Earth was a sphere, but Eratosthenes was able to go further and use sun angles at different locations to determine the actual circumference of the spherical Earth. He also determined that Earth is tilted on its axis. Eratosthenes wrote books about Earth, and correctly divided our world into polar, temperate, and tropical climate zones. He recognized that cities and countries at the same distance from the equator—what we now call latitude—had similar climates. He created a map that showed latitude lines running parallel to the equator, as well as lines that ran perpendicular to these lines of latitude; what we now recognize as lines of longitude.

Objectives

After studying this chapter and completing the exercises, you should be able to do each of the following tasks, using supporting terms and principles as necessary.

1. Read latitude and longitude values from a map, with a precision of minutes.
2. Explain why navigators have been able to determine latitude since ancient times, but it wasn't until the 1700s that navigators could accurately determine longitude at sea.
3. Describe how GPS receivers use satellites to determine locations on Earth.
4. Explain what map projections are and why they are necessary for the creation of maps.
5. Describe the shape of Earth.
6. Define remote sensing and give examples of how remote sensing is used to study Earth.
7. Distinguish between GPS and GIS.
8. Interpret features on a topographic map, especially contour lines.
9. Calculate gradients and percent slopes from a topographic map.

Vocabulary Terms

You should be able to define or describe each of these terms in a complete sentence or paragraph.

1. azimuthal projection
2. cartographer
3. cartography
4. conic projection
5. contour interval
6. contour line
7. coordinate system
8. cylindrical projection
9. echo sounding
10. electromagnetic spectrum
11. Geographic Information System (GIS)
12. Global Positioning System (GPS)
13. gradient
14. index contour line
15. latitude
16. longitude
17. map legend
18. map projection
19. map scale
20. oblate spheroid
21. percent slope
22. prime meridian
23. remote sensing
24. scale bar
25. topographic map

2.1 Locations on Earth

We do not have an original copy of the ancient world map made by Eratosthenes, and shown at the beginning of the chapter. The map shown in the image is a reconstruction based on written descriptions in the writings of Eratosthenes and other ancient scholars. But we know Eratosthenes had a good understanding of the size and shape of our planet, as well as of the distribution of climate and vegetation zones. He also introduced the general concepts of latitude and longitude, which have been refined in the centuries since then.

You may have already discovered that maps are fascinating. It is easy to pick up a map to look up a location, only to find yourself studying all the details for an hour or more! *Cartography* is the art and science of making maps, and a person who makes maps professionally is a *cartographer*. Cartography is an art, since a cartographer must have creativity and a sense of what makes a map pleasing to the eye. It is also a science, because it involves measurements, analysis of data about Earth, and an understanding of topics such as ecology, landforms, and electromagnetic radiation. Obviously, modern cartographers have many more tools available to them than Eratosthenes did, such as computers, satellites, and advanced surveying equipment.

2.1.1 Latitude and Longitude

Locations on Earth can be specified using a *coordinate system*, which is a system of using numbers to describe the location of a point on Earth. You are familiar with coordinate systems in mathematics, such as the Cartesian coordinate system, which plots an *x-y* ordered pair on a graph. The most widely used coordinate system for specifying locations on Earth is the latitude-longitude geographic coordinate system, in which locations are given as angles. *Latitude* is the angle between lines drawn from the center of Earth to the surface, as shown in Figure 2.1. The 0° line is the equator and other lines are measured north or south of the equator. Lines of latitude run parallel to the equator and to each other, so they are always the same distance apart. Because of this, lines of latitude are sometimes called *parallels*.

Navigators at sea have been able to determine latitude since ancient times. On a clear night in the Northern Hemisphere, it is fairly easy to determine the latitude by observing Polaris, the North Star. For an observer on Earth's surface, the angle between the lines of sight to the horizon and

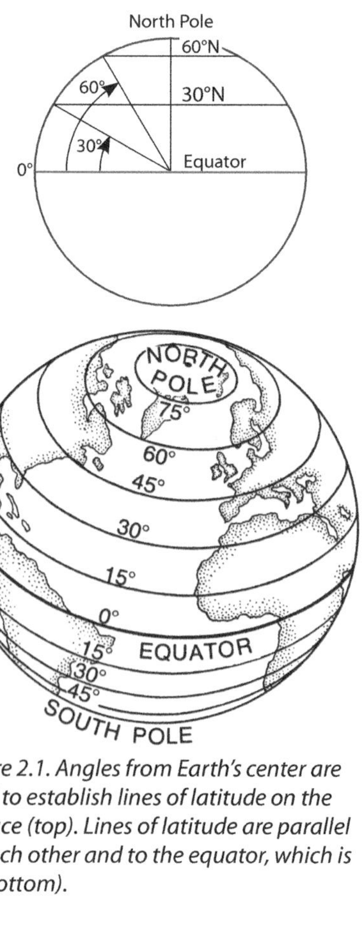

Figure 2.1. Angles from Earth's center are used to establish lines of latitude on the surface (top). Lines of latitude are parallel to each other and to the equator, which is 0° (bottom).

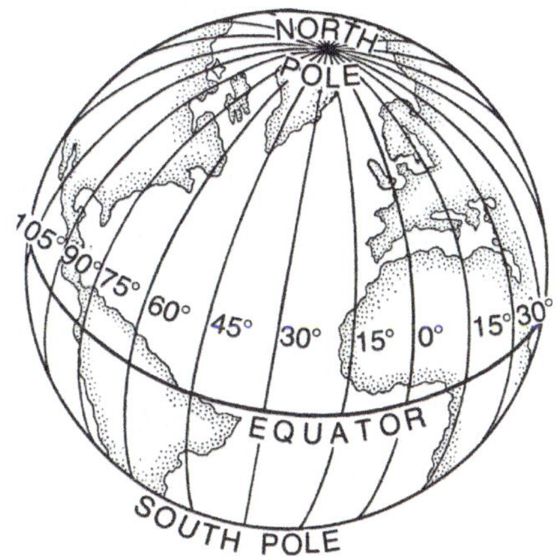

Figure 2.2. Lines of longitude. The prime meridian runs through Greenwich, England, and is designated as 0°.

to Polaris in the night sky is very close to the observer's latitude. Polaris is almost directly above Earth's North Pole, so if it is directly overhead, you know you are at 90°N (the North Pole). At a point half way between the equator and the North Pole, Polaris would be 45° above the horizon, and the latitude would be 45°N. There is no clearly visible star directly above the South Pole, so there is no commonly used pole star in the Southern Hemisphere.

Longitude is the angle east or west of a 0° line running from the North Pole to the South Pole. This 0° line is called the *prime meridian.* The lines of longitude, shown in Figure 2.2, are sometimes called *meridians.* Meridians are measured as angles east or west of the prime meridian, and longitude values range from 0° to 180°, for a total measurement of 360° around the entire planet.

In the past, many countries had their own prime meridians; there was a meridian of Paris, one of London, another of Washington D.C., and many others. In 1884, an agreement was reached at an international conference to recognize the meridian going through Greenwich, England to be the international standard prime meridian. The straight north-south boundaries between many of the western states in the United States were originally defined as being a certain number of degrees west of the Washington D.C. meridian, rather than being a certain longitude west of Greenwich.

Unlike lines of latitude, lines of longitude are not parallel, but all come together to a point at the poles. At the equator, one degree of latitude and one degree

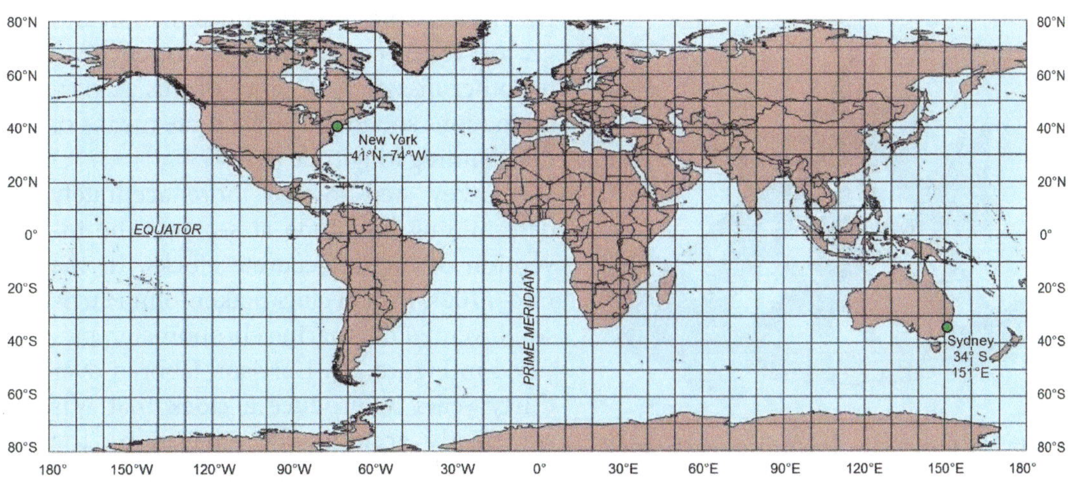

Figure 2.3. Latitude and longitude of New York City and Sydney, Australia.

Figure 2.4. One of Harrison's marine chronometers. The invention of accurate clocks in the 1700s enabled navigators to make accurate plots of their positions at sea for the first time.

of longitude cover very close to the same distance on Earth's surface. As one travels closer to a pole, the lines of longitude get progressively closer to one another.

Locations on Earth's surface are expressed with latitude first, then longitude. On Figure 2.3, you can see that New York City is located at 41°N, 74°W (stated as 41 degrees north, 74 degrees west), and Sydney, Australia is at 34°S, 151°E. On a global scale, this might be sufficient to locate a place, but on a local scale, latitude and longitude may be expressed with a much higher degree of precision. Each degree of latitude or longitude is divided into sixty minutes, and each minute is divided into sixty seconds. Minutes are represented by the symbol ′ and seconds by the symbol ″. This enables us to state locations more precisely. For example, the Statue of Liberty in New York City is at 40°41′21″N, 74°2′41″W.

Although mariners have known how to determine latitude at remote locations for many centuries, they did not have tools for accurately determining longitude until the mid-1700s. On land, people could use surveying techniques to measure distances and angles, and thus determine how far east or west a location was from a central meridian, but those methods could not be used at sea. A common method for determining locations at sea was by dead reckoning, which was based on the estimated course of a ship. A ship would leave from a known location, and the navigator would estimate how far it had travelled based on the speed of the ship through the water. Sometimes this worked well enough, and ships would meet their planned destinations. There were numerous instances, however, when ships or entire fleets would miss their destination and get lost at sea, crash into rocks, or encounter other disasters. Because of this, seafaring nations sought more accurate means of determining longitude.

Figure 2.5. A sextant is used to measure the angle between the horizon and the sun or other celestial body.

A key step toward more accurately determining longitude at sea was the invention of highly accurate clocks. In the mid-1700s, British clockmaker John Harrison invented the marine chronometer, pictured in Figure 2.4. Harrison labored over thirty years to produce a clock that was sufficiently accurate for navigation. The invention of the marine chronometer enabled navigators to use the positions of the sun, moon, planets, and stars at a specific time accurately to determine both latitude and longitude. The navigator would use a sextant, such as seen in Figure 2.5, to measure the angle between the horizon and an astronomical body—the sun by day or moon, planet, or star by night—and then consult books (called nautical almanacs) with printed tables of the positions of these bodies at various times. This would allow the navigator to determine a ship's position to within a few miles, which was usually sufficient to get the ship to its destination and avoid disasters.

Figure 2.6. A Global Positioning System (GPS) satellite, with solar panels to supply it with electricity.

2.1.2 Satellite Navigation Systems

Today, locations on Earth can be precisely determined by using signals from satellite navigation systems. The most commonly used satellite navigation system is the *Global Positioning System*, or GPS. A GPS satellite is shown in Figure 2.6. The Global Positioning System always has a minimum of 24 satellites orbiting Earth. These satellites continuously transmit radio signals that can be used to determine locations on Earth. Figure 2.7 shows how the satellites surround the planet. At any given moment, there are at least four satellites in the sky over any location. The radio signals sent by the satellites contain the time the signal was sent, and the position of the satellite when the signal was sent.

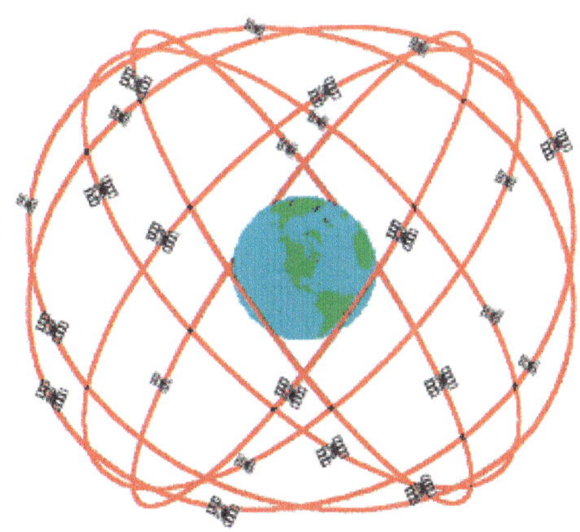

Figure 2.7. The 24 satellites of the Global Positioning System (GPS) provide worldwide service.

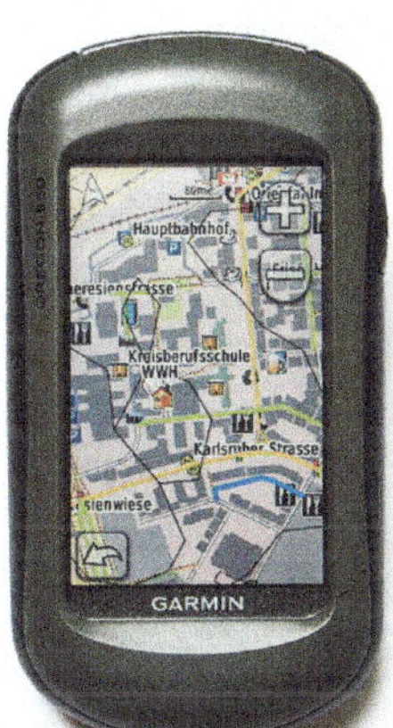

Figure 2.8. GPS receiver with a map showing the location of the receiver.

Figure 2.8 shows a typical hand-held GPS receiver. The receiver determines the amount of time it has taken for the signal to travel from each of the satellites. Since the signals travel at the speed of light (300,000 kilometers per second), the receivers must be able to measure time intervals with a precision of billionths of a second. Some of the satellites will be closer to the receiver and others will be farther away. If the receiver can calculate the distance to four different satellites with known positions, then the receiver can calculate its own position, as illustrated in Figure 2.9.

Satellite navigation now has hundreds of applications. These systems were originally developed for military use, but now are used for airplane, ship, and car navigation; surveying; studies of Earthquakes and volcanoes; tracking wildlife; and recreational uses such as backpacking. Many smart phones and tablet computers now have GPS receivers built into them. The Global Positioning System is operated by the United States, but there are similar systems operated by Russia and China, and other satellite navigation systems are being developed by other countries.

Figure 2.9. The spheres represent different distances from GPS satellites. The point at which all four spheres intersect is the location of the receiver.

Learning Check 2.1

1. Compare lines of latitude to lines of longitude, explaining how they are similar and how they are different.
2. Explain how Harrison's invention of the first marine chronometers enabled navigators to determine their longitude while at sea.
3. Describe how the GPS system is used.

2.2 Map Projections and the Shape of Earth

Imagine a balloon with a map of the United States drawn on it. If we were to pop that balloon, it would be a challenge to lay that map completely flat on a table. As shown in Figure 2.10, we would have to stretch and distort the rubber in order to get the United States, or any other portion of the balloon, to be flat.

Figure 2.10. The curved surface of the United States fits properly on the inflated balloon, but the popped balloon would have to be stretched to get the United States to fit on a flat surface.

This is the same problem cartographers have when they take regions on the roughly spherical Earth and make maps on flat surfaces, such as a sheet of paper or computer monitor. The true sizes, angles, and shapes of features on Earth—continents, oceans, countries, and so forth—need to be distorted in order to make a flat map. These distortions can be obvious on maps like those in Figure 2.11, which show large areas, but distortions exist even on maps of small areas, such as on a map of a city.

Figure 2.11. Both of these maps distort the size and shape of the United States and Canada. The only true portrayal of the size and shape of features on Earth's surface is on a globe.

2.2.1 Map Projections

A *map projection* is a method by which the curved surface of Earth is portrayed on a flat surface. To visualize how projections work, imagine drawing lines from the center of the globe through features on Earth's surface, and extending these lines onto a surface that can be laid flat. There are numerous projections in use, but most of them fall into a few categories.

A common type of projection is a *cylindrical projection*, which is created by extending features on Earth's surface onto a cylinder wrapped around Earth, as illustrated in Figure 2.12. There are several ways of doing this. A common example is the Mercator projection, which portrays all lines of latitude and longitude as parallel to each other. Because lines of longitude in reality meet at the poles, the Mercator projection distorts the sizes of features closer to the poles. If you look at a globe—on which features are portrayed in their true proportions—you will see that Greenland is considerably smaller than South America. On a Mercator projection, however, Greenland can look as large, or even larger than, South America, such as on the world map shown in Figure 2.13. The advantage of a Mercator projection, however, is that both Greenland and South America are preserved in roughly their true shapes. Another advantage is that straight lines drawn on a Mercator projection map follow a constant compass direction, and thus are easy to use for navigation. Mercator projections are frequently used for world maps, and are also often used for maps of smaller areas, such the topographic map shown in Figure 2.14.

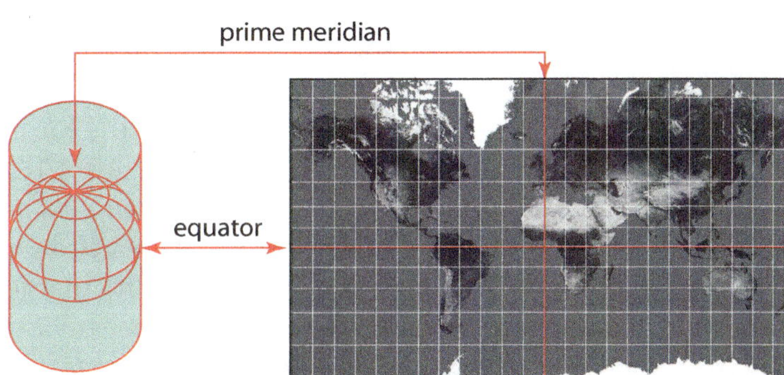

Figure 2.12. For a cylindrical projection, features on a globe are projected onto a cylinder that is wrapped around the globe.

A *conic projection* is created by extending features on Earth's curved surface onto a cone, as shown in Figure 2.15. This reduces distortion of features near the latitudes at which the imaginary

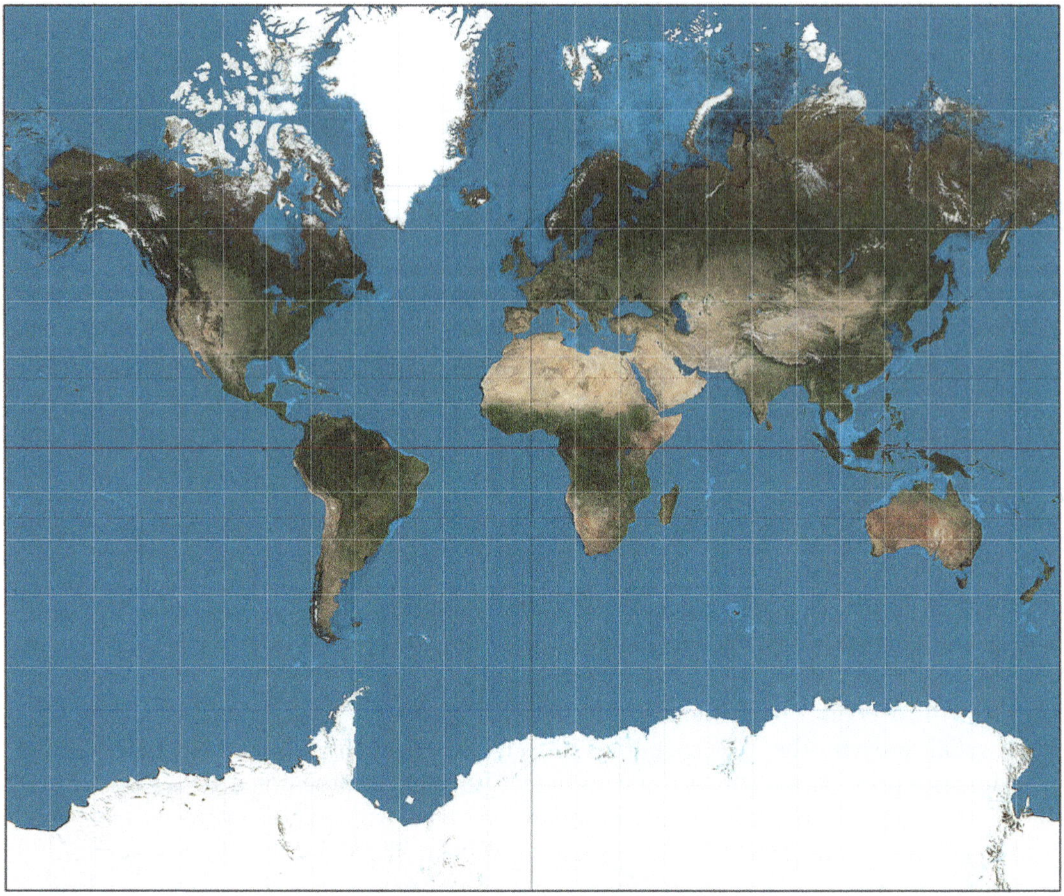

Figure 2.13. A world map on a Mercator projection, from 82° S to 82° N.

cone intersects Earth's surface. Maps of entire countries or continents are commonly drawn using conic projections. The projection is designed so there is a minimum amount of distortion of size or shape in the area of interest. Usually, the further one goes from the center of the map, the greater the distortion, as is apparent on the north and south edges of the map on Figure 2.16.

An *azimuthal projection*, shown in Figure 2.17, is created by projecting Earth's surface onto a plane that touches the planet at only

Figure 2.14. A topographic map of Mount Shasta, California, on a Mercator projection. For maps of small areas such as this, distortions due to the map projection are not significant for most purposes.

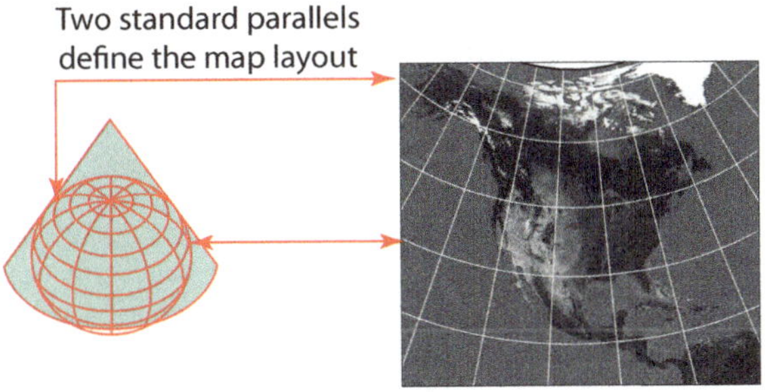

Figure 2.15. For a conic projection, features on a globe are projected onto a cone that is placed over a globe.

Figure 2.16. A map of North America on a conic projection. This preserves the size and shape of features near the central latitudes of the map, but distortions increase to the north and south.

one point. There is no distortion of features at this point, and the farther one gets from that point, the larger features become. The most common type of map that uses an azimuthal projection is one centered on one of Earth's poles. In Figure 2.18, the Arctic Ocean is portrayed much as it would be on a globe, with little distortion of size or shape. Features become increasingly distorted as the distance from the North Pole increases.

2.2.2 The Shape of Earth

There are many variations of each of these map projections, each with advantages and disadvantages. The only way to really portray Earth without significant distortions is on a globe. But a perfectly spherical globe would still have slight distortions. When I have been describing Earth, I have been careful to use phrases like "roughly spherical." In fact, Earth actually has an equatorial bulge, meaning that it is slightly flattened at the poles, and has a greater circumference when measured around the equator than when measured from pole to pole. A more accurate term for Earth's shape is *oblate spheroid*. An oblate spheroid, portrayed in Figure 2.19, is a flattened sphere where the

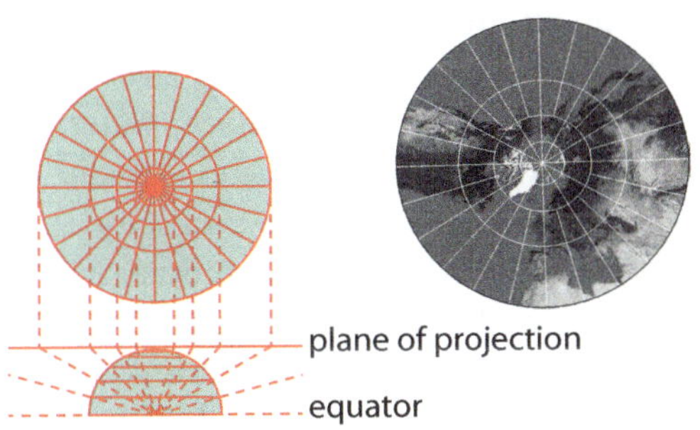

Figure 2.17. For an azimuthal projection, features on a globe are projected onto a plane that touches the globe at one point.

Figure 2.18. A map centered on the North Pole with an azimuthal projection. Features near the North Pole have little distortion of size or shape, but areas at a greater distance from the pole are noticeably stretched out.

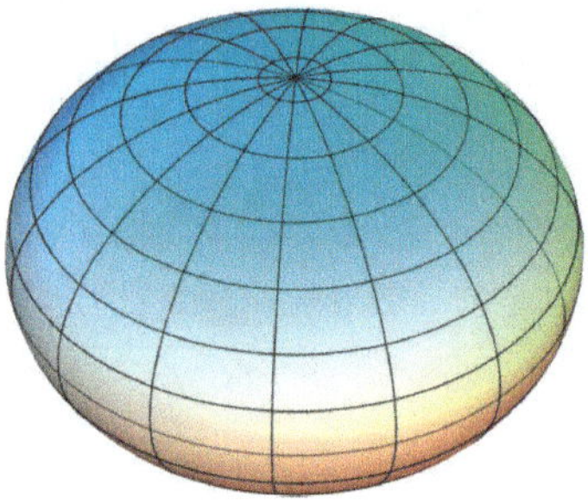

Figure 2.19. Rather than being a perfect sphere, Earth is slightly flattened at the poles and has a slight bulge at the equator. This shape is called an oblate spheroid. The amount of flattening is greatly exaggerated in this diagram.

equator has a greater radius than the measurement from the center to a pole. If gravity were the only force acting on material in Earth, it would be a sphere, but Earth's daily rotation on its axis causes it to flatten at the poles relative to the equator.

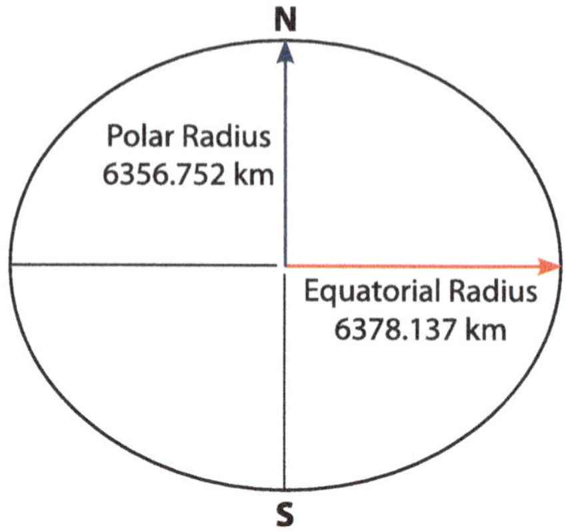

Figure 2.20. Earth's radius measured through the equator is slightly greater than its radius measured through the poles.

The difference between Earth's radius measured through the poles and through the equator is less than 1%, which is why Earth seems spherical. However, all cartographic and scientific measurements of Earth, such as latitude and longitude, are actually made by considering Earth as an oblate spheroid rather than as a sphere. The most commonly accepted measurements for Earth's radius are indicated in Figure 2.20.

Learning Check 2.2

1. Explain why it is impossible to make an accurate map of Earth on a flat sheet of paper without using a projection.
2. Describe the difference between cylindrical, conic, and azimuthal projections.
3. Describe the shape of Earth.

2.3 Remote Sensing

Remote sensing is the collection of information about Earth and the environment from a distance, usually by detecting and recording energy from the *electromagnetic spectrum* emitted from or reflected by Earth's surface, vegetation, water, or atmosphere. The electromagnetic spectrum is shown in Figure 2.21, and is made up of electromagnetic energy ranging from high-energy gamma rays and X-rays down to low-energy radio waves, with the visible light spectrum in the middle. All portions of the electromagnetic spectrum have been used in remote sensing.

Radio waves, on the left side of Figure 2.21, have long wavelengths but low energy. You can remember that radio waves have low energy when you think of how many radio signals are going through you right now—from every radio and TV station in your area, from cell phones, and from other sources—with no apparent effect.

X-rays and gamma rays are on the other end of the spectrum. They have very short wavelengths—measured in billionths or trillionths of a meter—but very high energy.

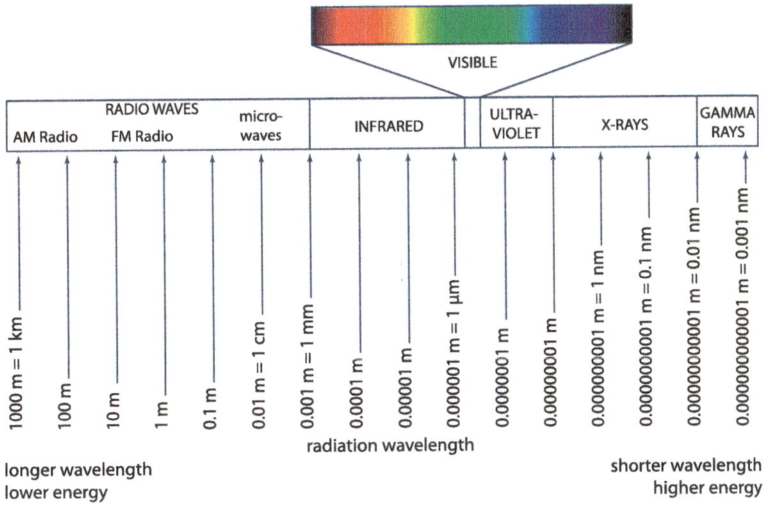

Figure 2.21. The electromagnetic spectrum.

Data from remote sensing is often presented in the form of pictures referred to as *imagery* (or sometimes referred to as *aerial or satellite photography*). If you have used internet sites or apps on mobile devices to view an overhead picture of your house, school, or other location, you have viewed remote-sensing imagery.

Figure 2.22. Before airplanes became common for aerial reconnaissance, attempts were made to use pigeons as aerial photographers.

Some of the earliest aerial photography was taken using cameras mounted to balloons, kites, and even pigeons, as shown in Figure 2.22. Today, remote-sensing data are typically collected from above using airplanes or satellites, such as the NASA Landsat satellite shown in Figure 2.23. Much of this technology was initially invented for military and intelligence purposes, leading to the development of what are commonly referred to as "spy satellites," but many other uses have now been developed.

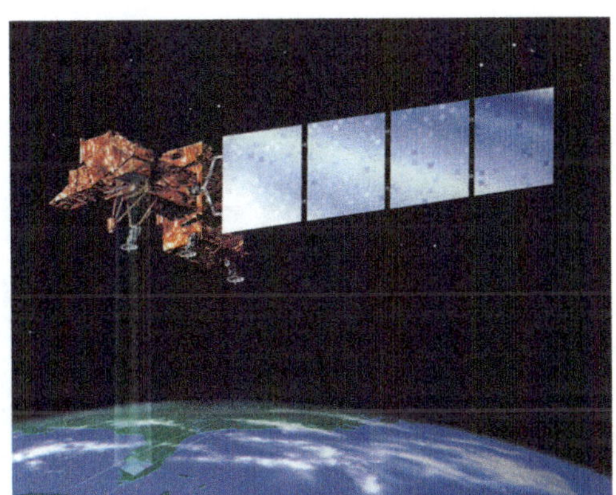
Figure 2.23. Landsat satellites capture images of the entire Earth once every 16 days.

Remote sensing satellites have provided regular coverage of most of Earth for over forty years. Imagery from these satellites are used to trace changes to forests, farmland, cities, water bodies, and many others. In addition to using visible light, many of these satellites record infrared radiation, which means that the satellite detects wavelengths of light that the human eye cannot. This enables the sensor to distinguish

Figure 2.24. Satellite imagery can be used to quickly analyze damage to forests due to wildfires. In this image, red is healthy vegetation, gray is bare ground or man-made features, and dark brown is the burn scar from a wildfire.

Figure 2.25. Remote sensing images from weather satellites give detailed and timely information about Earth's atmosphere. In this image centered on North America, the highest clouds, which are associated with the tops of strong storm systems, are red and orange.

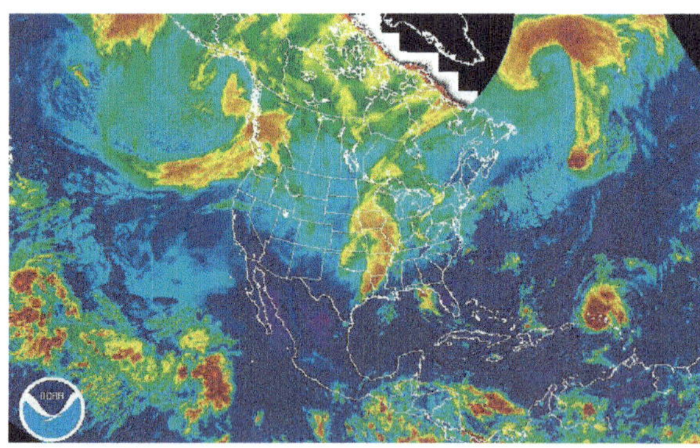

Figure 2.26. Echo sounding uses sound waves transmitted through water to determine water depth.

changes in vegetation and soil moisture that are too subtle to be seen with visible light. Figure 2.24 shows the burn scar from a large forest fire in Colorado. Healthy trees are shown with red to emphasize that these data were collected using infrared wavelengths, in addition to visible light.

Infrared wavelengths are also useful for geological investigations, such as searching for mineral deposits.

Remote-sensing images are important for many uses in oceanography and meteorology. In the past, weather observations were limited to those made at manned weather stations, and there were large parts of Earth where storms could develop without anyone knowing about them. Now we have a continuous global picture of what is happening in the atmosphere. The first weather satellites sent back images that enabled forecasters to see the extent of cloud cover, but not much more. Instruments on modern satellites detect the temperatures of cloud tops, as shown in Figure 2.25, and measure wind speeds, precipitation, lighting strikes, concentrations of various pollutants, and much more. Since Earth's land surface, oceans, and atmosphere emit infrared radiation at night as well as during the day, weather satellites even gather remote-sensing data in the dark.

Figure 2.27. A map of a submarine volcano in the Aegean Sea, with elevations determined using echo sounding. The main crater is about 1.5 kilometers across (about 1 mile). The deepest part of the crater (purple) is about 500 meters below sea level, and the shallowest part of the crater rim (white and dark red) is about 10 meters below the ocean surface.

Not all remote sensing is done with electromagnetic radiation. An example of a different technology is *echo sounding*, which uses sound waves rather than electromagnetic radiation to determine the depth of water in oceans or lakes. By measuring the time required for the sound waves to travel from the ship to the sea floor and back to microphones in the water, oceanographers are able to determine the depth of the water, as illustrated in Figure 2.26. It takes a large amount of echo-sounding data to map even a small part of the ocean floor, such as the map of a submarine volcano depicted in Figure 2.27. Sound waves are also transmitted into the solid Earth to determine the structure and layering of rocks in Earth's

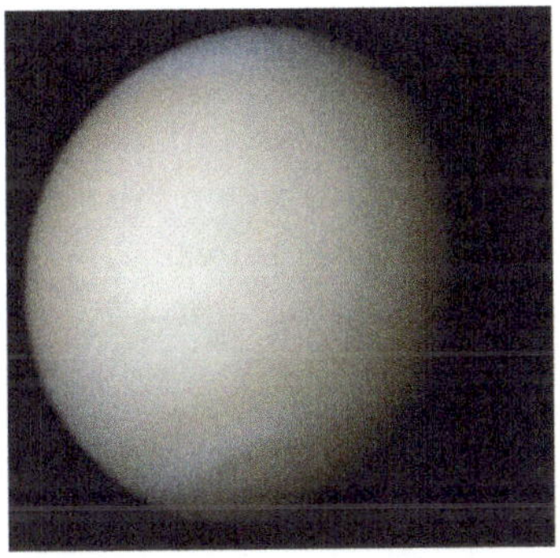

Figure 2.28. Because of its thick layers of clouds, it is impossible to see the surface of Venus from space using visible light.

Figure 2.29. Use of radar on NASA's Magellan probe allowed scientists to make a complete map of Venus. The colors are added by computer software. As in Figure 2.27, blue areas are low in elevation, and red and white areas are high. The surface of Venus is extremely hot and dry, with an average surface temperature of 462°C (863°F).

crust and deeper. This technique is commonly used by geologists exploring for petroleum deposits.

Probes sent to other parts of the solar system also use remote sensing to gather information, using the same principles that are used for observing Earth. We would know very little about the surfaces of Mercury, Venus, Mars, or the moons of the outer planets if it were not for remote sensing. For example, Venus is enveloped by a thick layer of clouds, as shown in Figure 2.28, and its surface is never visible from space. The surface of Venus, however, has been mapped in great detail using radar mounted on the Magellan probe. Radar determines distances using radio waves. The transmitter sends out radio waves which bounce off the surface of a planet and back to a receiver on the satellite. The Magellan probe was able to gather data to make an elevation map of almost the entire surface of Venus, as shown in Figure 2.29.

Learning Check 2.3

1. List the six regions of the electromagnetic spectrum, from low energy to high energy.
2. Describe a way that remote sensing is used to study each of the four Earth systems (lithosphere, hydrosphere, atmosphere, and biosphere).

2.4 Mapping Earth

A map is a model of the world, or of a part of the world, showing natural and man-made features. Creating a map involves mathematics, computer skills, geographic knowledge, and a lot of patience. But a good map is not just a technical document; it should also be pleasing to the eye. The "art" side of cartography involves making good decisions about colors, text placement, and map symbols, and these choices can make the difference between creating a good map and one that is difficult to read or even misleading. In the past, maps were usually printed on paper, but now people are increasingly turning to digital maps—on computer screens or mobile devices—on a day-to-day basis.

There are many types of maps, such as road maps and political maps. Maps are used in Earth sciences to present information about all sorts of topics, ranging from weather maps and maps of ocean surface temperatures, to maps showing the topography of mountain ranges or of the sea floor. You have heard the phrase, "A picture is worth a thousand words." The same concept applies to cartography: "A map is worth a thousand words." The preparation and use of maps can powerfully and quickly communicate concepts to both scientific and general audiences. Consider a map such as the one in Figure 2.30, which shows the average wind speed 80 meters above the ground for the United States. This map was created as part of an effort to identify the best locations for placing wind power turbines, with the purple areas being those with the highest average wind speeds, and the yellow and green areas being the areas with lowest average wind speeds. You can quickly look

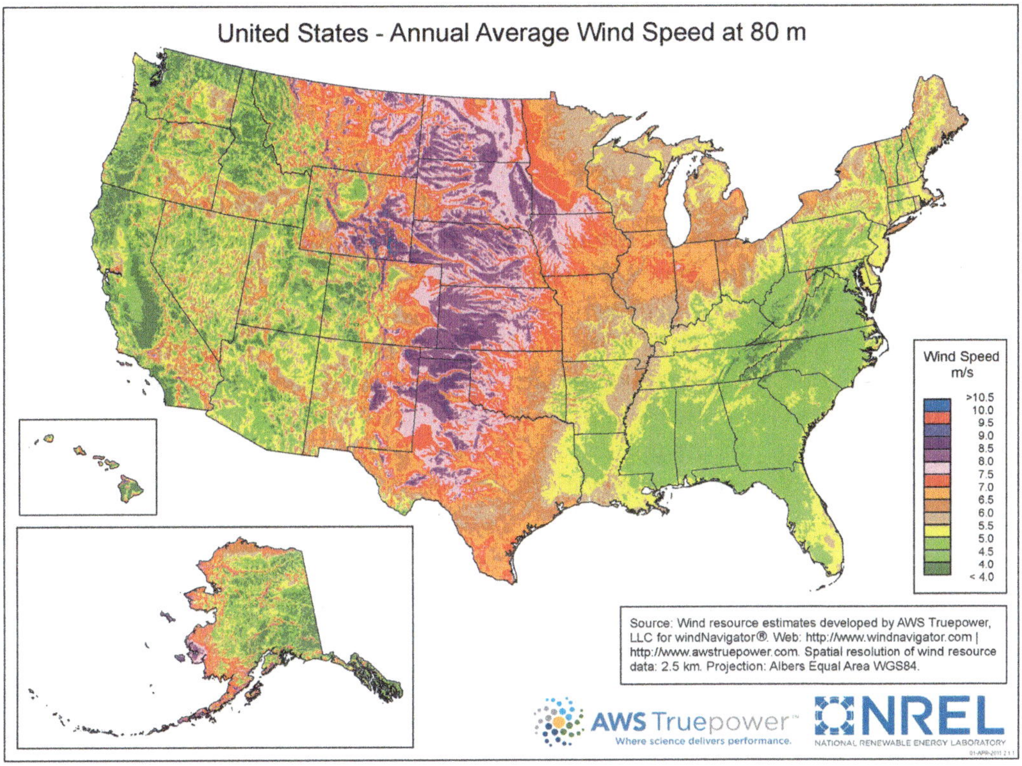

Figure 2.30. Map showing average wind speed, in meters per second, at 80 meters above the ground.

at the map and see that the areas with strongest winds are in the Great Plains, from northwest Texas up through North Dakota.

2.4.1 Geographic Information Systems

All maps are prepared using map projections, because the features portrayed on the map must be distorted in order to fit on a flat surface. In the past, this was a laborious process, but it is now done quickly by computers. Almost all maps, whether paper or digital, are now made using Geographic Information Systems. A *Geographic Information System* (or GIS) is computer software that analyzes data about Earth, and then presents it in the form of maps. The data contained in the GIS may include imagery, elevation data, as well as points, lines, and polygons that represent natural and man-made features—such as towns, roads, streams, lakes, and political boundaries. Using GIS software, maps that would have taken months to make in the past are often produced in a matter of hours or days.

There is much more to a GIS than just making maps. Cartographers or other GIS users also compare different sets of data to help make decisions. For example, a forester might use the GIS to compare soil types, elevations, and sun angles in order to decide the best places to plant tree seedlings after a forest fire. Again, this would have been a very time-consuming task before the advent of GIS technology.

2.4.2 Topographic Maps

One important type of map in Earth sciences is the *topographic map*. A topographic map represents the surface of Earth, showing elevations in a measurable fashion such as by contour lines or elevation tints (the color bands used back in Figure 2.27). Topographic maps also show other natural and man-made features, such as streams, lakes, forests, roads, and buildings. They are used by geologists, foresters, wildlife biologists, engineers, planners, the military, and for outdoor recreation. In the United States, detailed topographic maps have been made for the entire nation by government agencies such as the U.S. Geological Survey and the U.S. Forest Service. A section from a U.S. Geological Survey topographic map is shown in Figure 2.31. The topographic maps made by these agencies are referred to as quadrangles. Most other countries have similar topographic mapping programs.

Most topographic maps portray elevation using *contour lines*, which are lines that connect points of equal elevation. The brown lines in Figure 2.31 are contour lines. Each point along the 900-foot contour has an elevation of 900 feet above

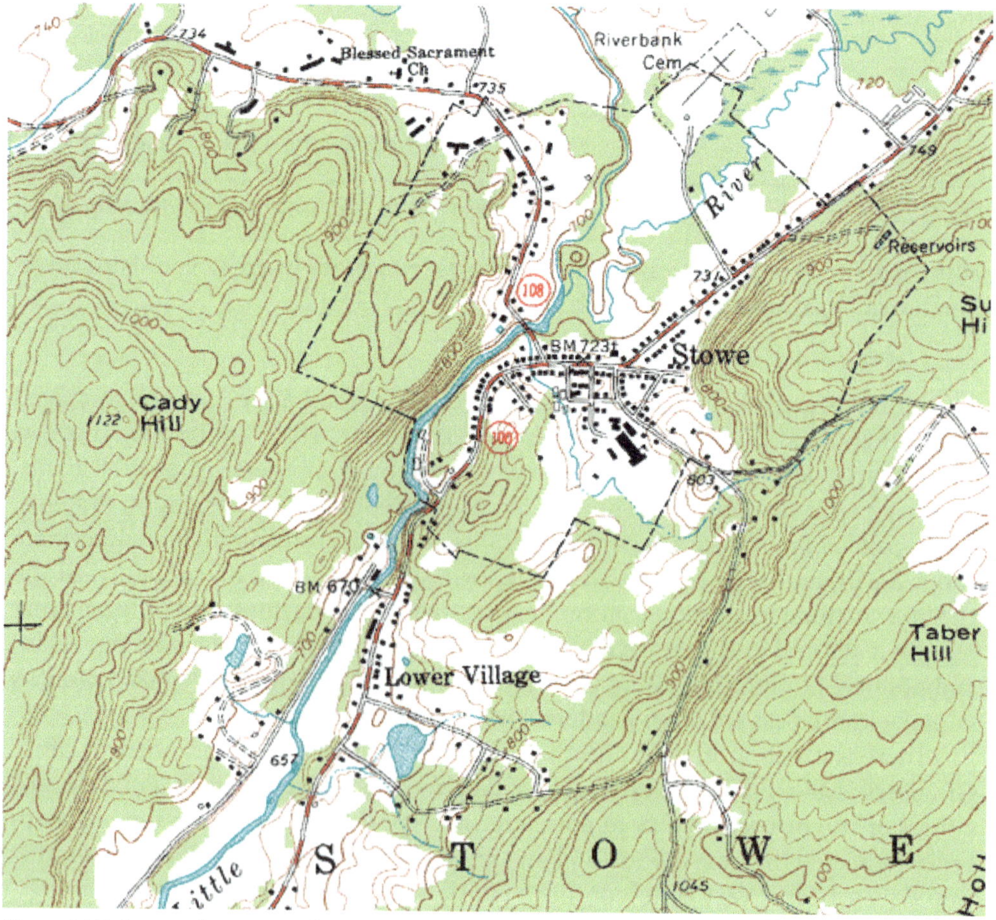

Figure 2.31. A part of a topographic map, showing roads, buildings, water bodies, trees, and elevations.

sea level, and each point along the 1,100-foot contour has an elevation of 1,100 feet above sea level. The shapes of the contour lines on the map reflect the shape of the land surface. The starting point for measuring elevation is sea level, which is assigned a value of zero. Most topographic maps in the United States show elevations in feet, but military topographic maps and topographic maps produced by other countries have elevations in meters.

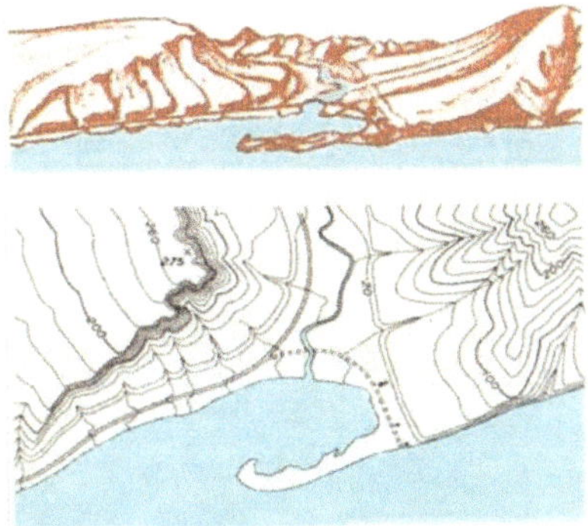

Figure 2.32. The relationship between landscape and contour lines.

Figure 2.32 illustrates the relationship between contour lines and landscape. In places where the land is relatively level, such as along the stream in the center of the diagram, the contour lines are far apart. Where the slope is steep, the contour lines are close to each other.

There are four basic principles associated with contour lines, illustrated in Figures 2.33, 2.34, and 2.35:

1. The first of these has just been stated: where contour lines are far apart, the slope is gentle; where they are close together, the slope is steep.
2. Contour lines never cross each other. On some topographic maps, they can touch each other where there are cliffs, but they do not actually cross each other.
3. Where a contour line crosses a stream on a topographic map, the contour line will be V-shaped, with the V pointing upstream, or to areas with higher elevation.
4. Contour lines form closed loops around elevated areas (e.g., hills) and depressed areas (e.g., craters). For example, a hill on a topographic map looks like a series of concentric, irregular loops. Often, the contour lines extend beyond the edges of an individual map so you don't always see the complete closed loop.

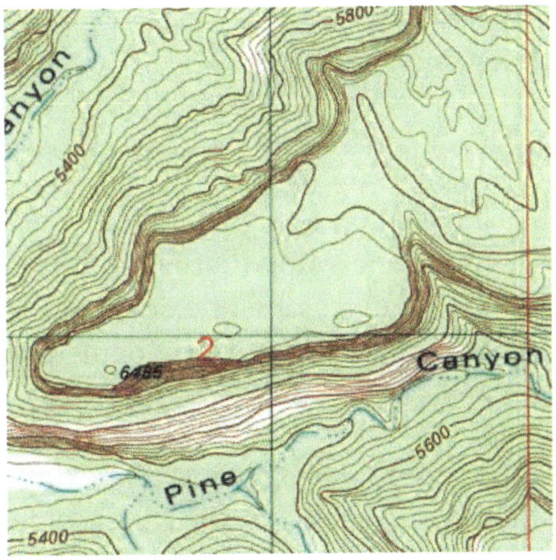

Figure 2.33. Principles 1 and 2—Contour lines are far apart in flat areas, and close to each other in steep areas, but contour lines never cross.

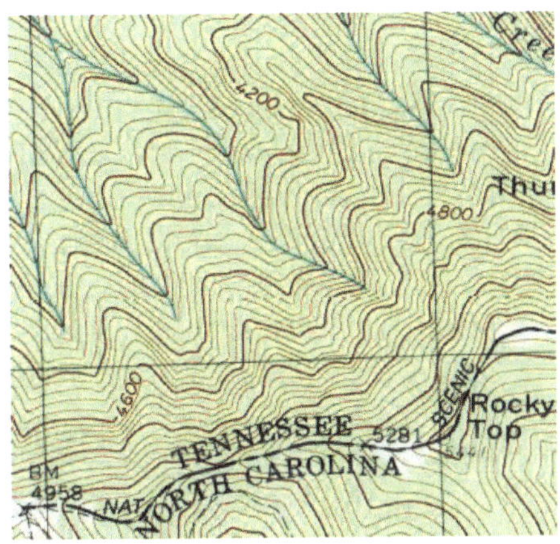

Figure 2.34. Principle 3—Contour lines make a V where they cross streams, with the V pointing upstream.

The *contour interval* of a topographic map is the difference in elevation between adjacent contour lines. If the contour interval is 20 feet, then contour lines will be shown for every mul-

Figure 2.35. Principle 4—Contour lines form closed loops around hills.

tiple of 20 feet, such as at 100, 120, 140, and 160 feet. For flat areas, there is not a great elevation difference from one part of a map to another, and a topographic map of an area of plains might have a contour interval of only 10 feet. Most hilly areas can be sufficiently portrayed with a 20-foot contour interval, and mountainous areas typically have contour intervals of 40 feet. If a mountainous area were portrayed with a 10-foot contour interval, the contours would be so close to each other that they would blend together.

In order to make a map easier to read, every fourth or fifth contour line, known as an *index contour line*, is drawn with a heavier line weight, and usually given a label indicating its value. If the contour interval on a map is 20 feet, then every fifth contour line has a value that is a multiple of 100 feet, and is designated as an index contour line. Contour interval and index contours are depicted in Figure 2.36.

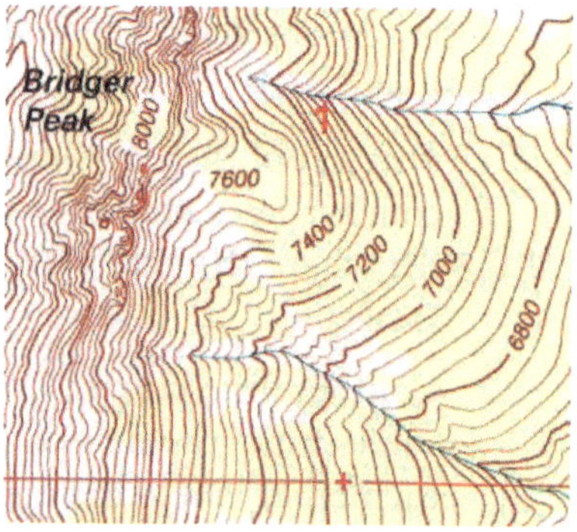

Figure 2.36. This map has a contour interval of 40 feet, with index contours every 200 feet.

2.4.3 Map Margin Information

Some of the most important features of a topographic map, or on other types of maps, are not found in the main part of the map but in the margins around the map's edge. The margin gives information about the location of the map, the publisher, the date, and accuracy of the map. One important element of the map margin is the *map scale*. The map scale is the ratio between a length on a map and the corresponding horizontal distance on the ground. For example, if a map has a scale of 1:100,000, it means that features shown on the map are one-100,000th of their true size. It also means that one unit of measurement on the map represents 100,000 of those same units on Earth. In this case, 1 cm on the map would represent 100,000 cm on the ground. This is a convenient scale, because it means that 1 cm on the map also represents 1 km on the ground. (There are 100,000 cm in 1 km.) Another way of expressing map scale is with a *scale bar*. A scale bar graphically represents the scale of the map, and consists of a line with marks like a ruler. The map scale and scale bar for a U.S. Geological Survey topographic map are shown in Figure 2.37.

Another important part of the map margin is the *legend*, as shown in Figure 2.38. A map legend is a key that shows what each symbol on the map represents. Many symbols on a map are designed to look like the features they symbolize. For example, trees are shown with a green tint and water bodies are depicted with blue. Many oth-

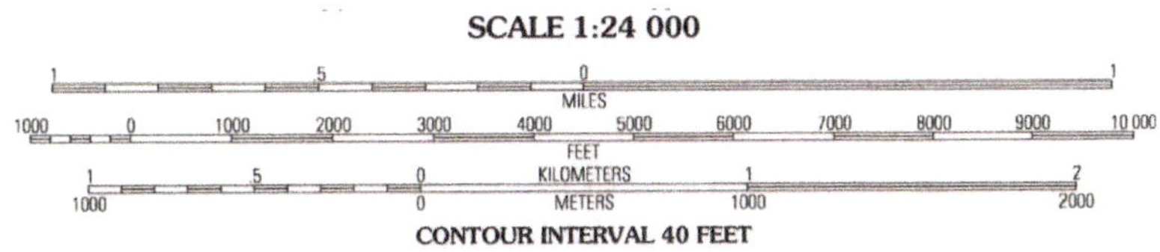

Figure 2.37. Scale bar and map scale from a topographic map.

er symbols, however, are not the same color as the features they represent. Man-made features, such as buildings, are usually black, no matter what color they really are.

2.4.4 Gradient and Percent Slope

We can speak of a slope of a geographic feature, such as a hillside, stream, or road, as being gentle or steep, or we can actually measure a slope to determine its *gradient*. Gradient is a mathematical measurement of the rate of change of the elevation of Earth's surface over a given horizontal distance. In algebra, the slope of a line is the ratio of "rise" over "run," and the gradient of Earth's surface is measured in a similar way. One way to express gradient is as the rise or drop in elevation per horizontal distance. For instance, if a stream drops in elevation 50 feet in 2.0 miles, its gradient can be calculated as

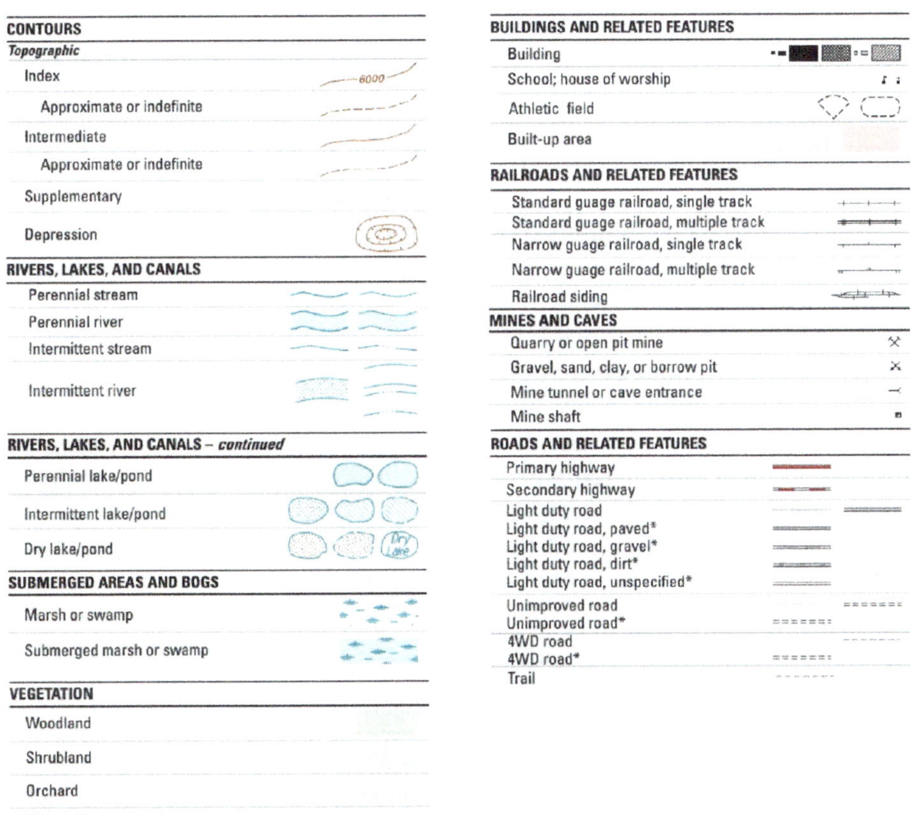

Figure 2.38. Simplified legend for U.S. Geological Survey topographic maps.

$$\text{gradient} = \frac{\text{rise (or drop)}}{\text{run}} = \frac{50\ \text{ft}}{2\ \text{mi}} = 25\ \text{ft/mi}$$

When expressed this way, the units of gradient will always be distance of rise or drop per distance of run, such as feet per mile or meters per kilometer.

Gradient is often presented as a *percent slope*. Percent slope is the rise over run of Earth's surface multiplied by 100%, as illustrated in Figure 2.39. When calculating percent slope, the units must be the same for both the rise and run, so the calculation uses an equation such as

$$\text{percent slope} = \frac{\text{rise feet}}{\text{run feet}} \times 100\%$$

Let's calculate the percent slope for the same stream. The stream drops 50 feet in 2.0 miles, but we must first convert the miles to feet. There are 5,280 feet in a mile, so the conversion is as follows:

$$2\ \text{mi} \cdot \frac{5280\ \text{ft}}{1\ \text{mi}} = 10{,}560\ \text{ft}$$

The stream's percent slope can now be calculated.

$$\text{percent slope} = \frac{\text{rise feet}}{\text{run feet}} \cdot 100\% = \frac{50}{10{,}560} \cdot 100\% = 0.47\%$$

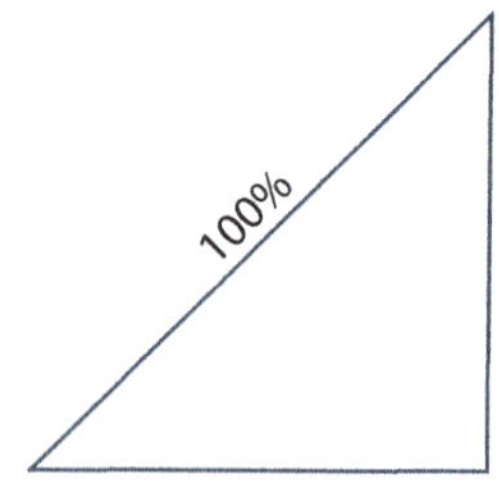

run = 100
percent slope
= (rise/run) × 100%
= (100 ÷ 100) × 100%
= 100%

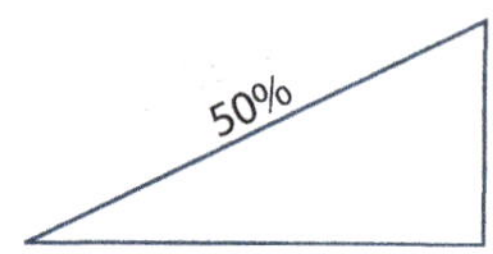

run = 100
percent slope
= (rise/run) × 100%
= (50 ÷ 100) × 100%
= 50%

Figure 2.39. Calculating percent slope.

Note that because there are units of feet in both the numerator and denominator of this equation, the units cancel out and the units are percent; the answer is 0.47 percent, not 0.47 feet per foot.

Gradient can be determined by interpretation of contour lines on topographic maps. To calculate the gradient of a hillside, for example, one would calculate the rise by determining the difference between the elevation for the highest contour line and the lowest contour line, and the run by using the map scale to measure the horizontal distance between those points. A sample calculation based on a topographic map is shown in Figure 2.40.

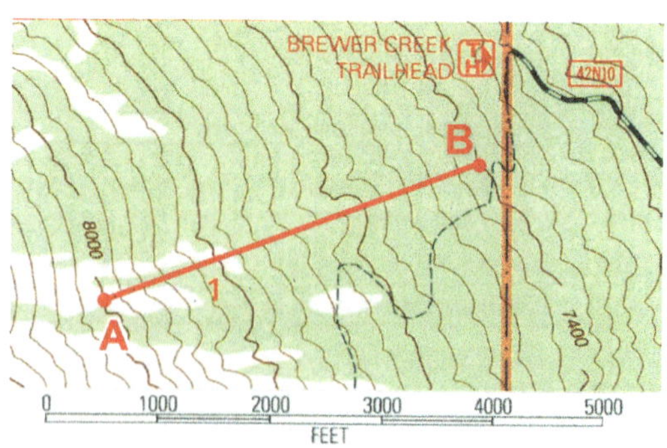

Figure 2.40. Point A has an elevation of 8000 feet, and point B has an elevation of 7400 feet, which gives a rise of 600 feet. The horizontal distance between A and B is 3600 feet. The percent slope is (rise/run) × 100% = (600 ft/3600 ft) × 100 % = 16.7%.

Learning Check 2.4

1. Explain what a Geographic Information System is.
2. Explain how contours are used to portray Earth's landscape on a topographic map.
3. What are the four basic principles of contour lines?
4. Explain what is meant by the term *map scale*.

Chapter 2 Exercises

Answer each of the questions below as completely as you can. Write your responses in complete sentences unless instructed otherwise.

1. Use the map of the United States below to determine the name of the cities with the following coordinates:

a. 40°N, 105°W	b. 42°N, 88°W	c. 26°N, 80°W	d. 46°N, 123°W
e. 41°N, 74°W	f. 34°N, 118°W	g. 42°N, 71°W	h. 30°N, 95°W

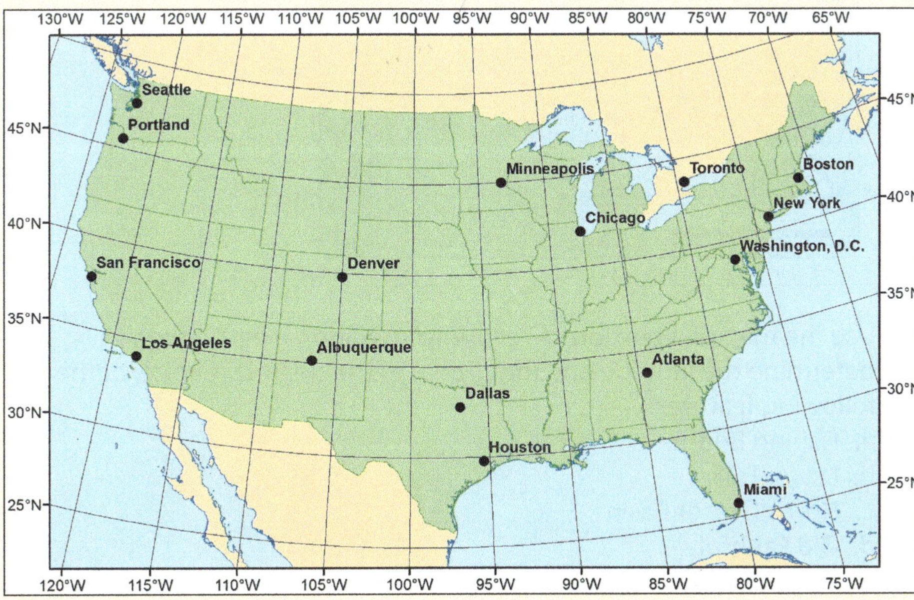

2. Use the map of the United States above to determine the latitude and longitude, to the nearest degree, of the following cities:

a. Albuquerque, New Mexico	b. Atlanta, Georgia
c. Dallas, Texas	d. Minneapolis, Minnesota
e. San Francisco, California	f. Seattle, Washington
g. Toronto, Ontario, Canada	h. Washington, D.C.

3. Use the map of a portion of Yellowstone National Park below to determine the names of the features with the following coordinates:
   a. 44°59′N, 110°42′W
   b. 44°48′N, 110°27′W
   c. 44°38′N, 110°27′W
   d. 44°32′N, 110°50′W
   e. 44°25′N, 110°34′W

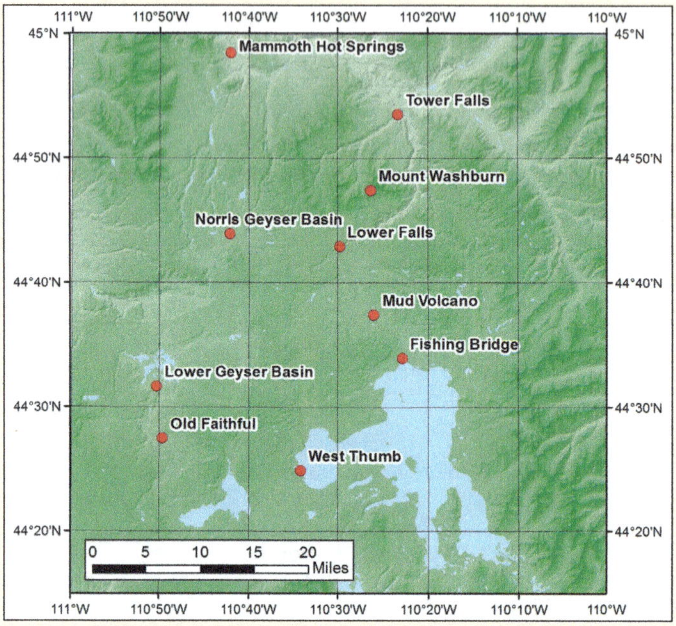

4. Use the map of a portion of Yellowstone National Park above to determine the latitude and longitude, to the nearest minute, of the following features:
   a. Fishing Bridge
   b. Lower Falls
   c. Norris Geyser Basin
   d. Old Faithful
   e. Tower Falls
5. Illustrate how features on a globe can be projected onto a cylinder, cone, and plane.
6. Describe the shape of Earth, and explain why it is not spherical.
7. Describe an example of remote sensing using one of the non-visible parts of the electromagnetic spectrum.
8. What are some advantages of using satellites to observe Earth's weather?
9. Explain why contour lines are far apart in flat areas but close together

in steep areas.

10. Use the topographic map of High Top Mountain below to answer the following questions (elevation values are in feet).
    a. What is the contour interval of this map?
    b. What is the interval between index contours?
    c. Estimate an elevation value for the x at the top of High Top Mountain.
    d. What are the gradients between A—B and C—D, measured in feet per mile?
    e. What are the percent slopes between A—B and C—D?

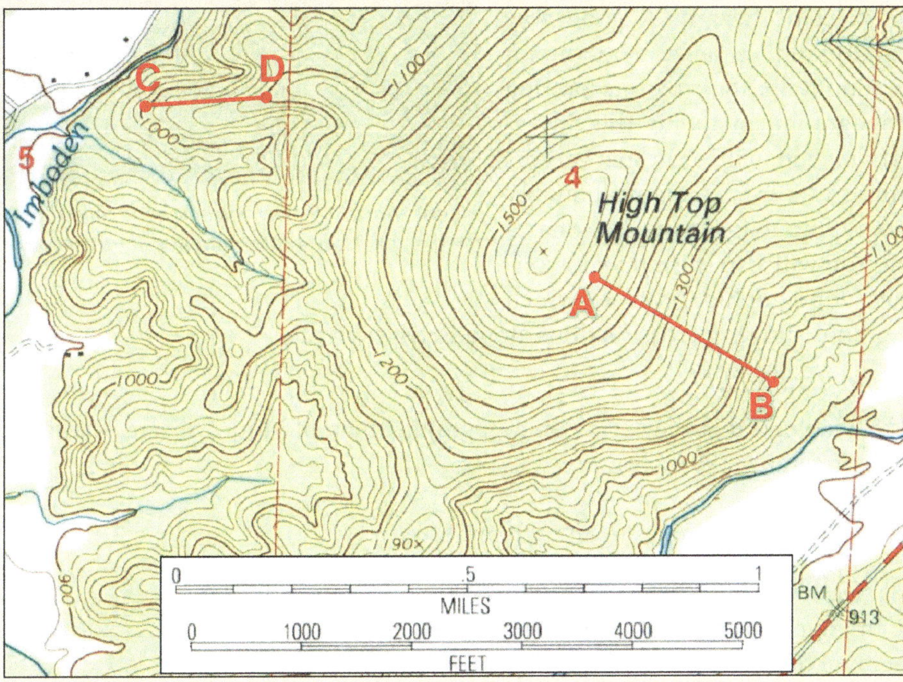

Answers

1. a. Denver b. Chicago c. Miami d. Portland e. New York f. Los Angeles g. Boston h. Houston

2. a. 35°N, 107°W b. 34°N, 84°W c. 33°N, 97°W d. 45°N, 93°W e. 38°N, 122°W f. 48°N, 122°W g. 44°N, 79°W h. 39°N, 77°W

3. Mammoth Hot Springs b. Mount Washburn c. Mud Volcano d. Lower Geyser Basin e. West Thumb

4. a. 44°34′N, 110°23′W b. 44°43′N, 110°30′W c. 44°44′N, 110°42′W d. 44°28′N, 110°50′W e. 44°54′N, 110°23′W

10. a. 20 ft b. 100 ft c. between 1580 and 1600 ft, e.g., 1590 ft d. 140 ft/mi, 48 ft/mi e. 26%, 9%

Experimental Investigation 1: Interpreting Topographic Maps

Overview

Mount Shasta is a volcano in northern California, and is the second tallest volcano in the Cascade Range of western North America. Mt. Shasta last erupted somewhere between 200 and 300 years ago, and is considered to be a considerable volcanic hazard. A good way to study and visualize a mountain like Mount Shasta is by interpreting a topographic map. In this investigation, you will examine various features of a U.S. Geological Survey topographic map of the Mt. Shasta area. You will also learn how to construct topographic profiles, which are graphs showing cross sections along Earth's surface.

Mount Shasta. A newer volcanic cone called Shastina is forming on its west side (right side in this image).

Basic Materials List

- U.S. Geological Survey 1:24,000 topographic map of Mount Shasta, California, preferably 1998 edition.
- Ruler
- Graph paper (5 squares to an inch preferred)

Part 1—Topographic Map Interpretation

In your Lab Journal, record the latitude and longitude of each corner of the map, and draw a sketch of the map with labels as illustrated to the right. Then address the following questions:

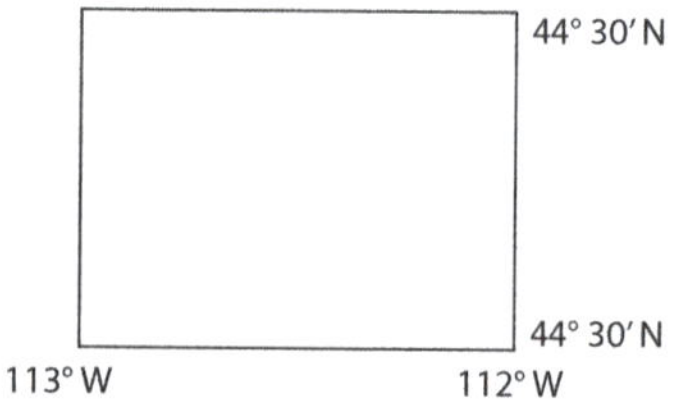

1. What is the scale of this map? What does that mean?
2. What is the distance, in a straight line, between the summit of Mt. Shasta and Clarence King Lake? Give your answer in miles and meters.
3. What is the contour interval of this map? What does that mean? What is the interval between index contours?
4. What is the elevation of the summit of Mt. Shasta? How do you know? What is this value in meters?

5. What is the elevation of Sisson Lake? How do you know? What is this value in meters?
6. What is the elevation difference between the top of Bolam glacier and the bottom of Bolam Glacier? What is the overall gradient of Bolam Glacier in feet per mile? What is the overall percent slope of Bolam Glacier?
7. What is the gradient along Road 43N21 in the northwest corner of the map? How does this compare to the gradient of Bolam Glacier? Describe how the contour lines relate to the gradients at these two locations.
8. The trail heading south from the Bolam trailhead (at the end of road 43N21 in Question 8) does not follow a straight line. Why not?
9. Where is the lowest elevation on the map, and what is the elevation at this location?
10. Compare the size of the glaciers (blue contour lines) on the north side of Mt. Shasta with the size of the glaciers on the south side of Mt. Shasta. Form a hypothesis as to why there is a difference in the areas of these glaciers.

Part 2—Constructing a Topographic Profile

A *topographic profile* is a cross section through Earth's surface along a line. One can think of a topographic profile as a graph of what Earth's surface would look like if one were to take a slice out of Earth's crust. The first step in constructing a topographic profile is to lay a strip of paper along the profile line, and transfer contour values to the paper, as shown below.

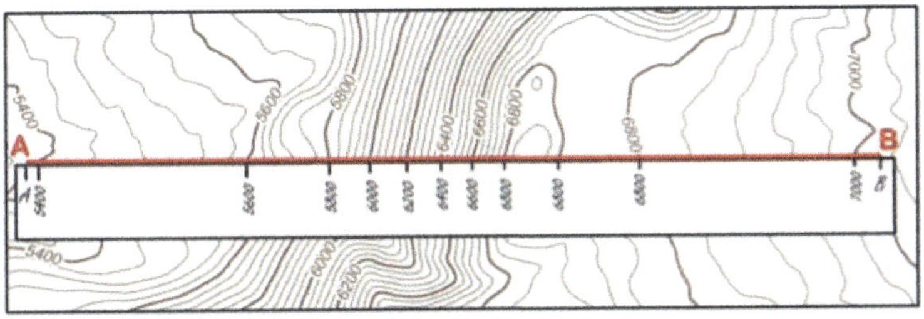

The second step is to take the strip of paper and transfer the values to a graph on graph paper, as shown by the black dots below. For the vertical scale on your graph, use the same scale as shown on the map.

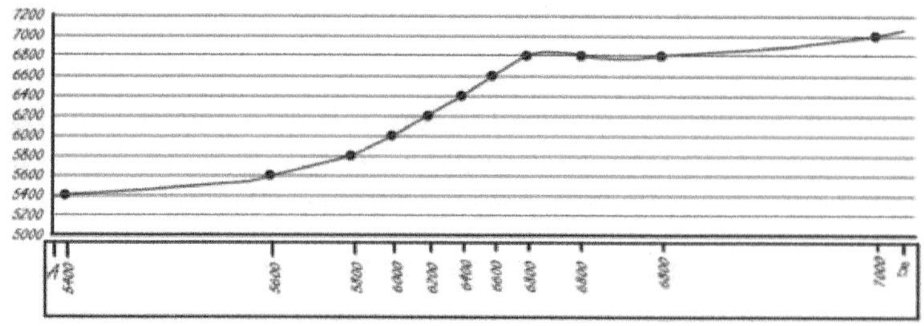

On a 1:24,000 topographic map, one inch represents 24,000 inches (2,000 ft) on Earth's surface. In the topographic profile above, the horizontal and vertical scales are both 1:24,000, so 1 inch in either the horizontal or vertical directions represents 2,000 ft. Since the scales on the horizontal and vertical axes of the graph are the same, the graph visually represents the true slope of Earth's surface. In some cases, such as in flat areas, it is preferable to have the vertical scale larger than the horizontal scale in order to emphasize subtle variations in the topography. This is called *vertical exaggeration.*

Finally, draw a smooth curve to represent Earth's surface along the profile line. Note that on the topographic profile, the curve is steeper where contour lines are closer together; the slope is more gentle where contour lines are further apart.

Construct topographic profiles from A to B for the three topographic maps shown on the next page.

Map 1: McCloud, California

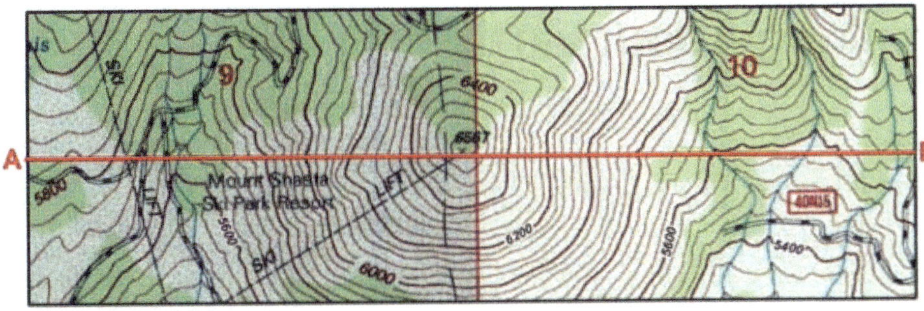

1:24,000 (One inch = 2,000 feet)
Contour interval 40 feet

0 1,000 2,000 3,000 4,000 Feet

Map 2: Rileyville, Virginia

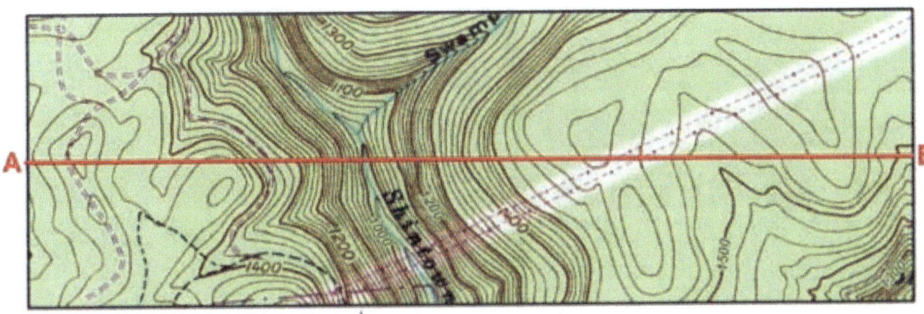

1:24,000 (One inch = 2,000 feet)
Contour interval 20 feet

0 1,000 2,000 3,000 4,000 Feet

Map 3: Renova West, Pennsylvania

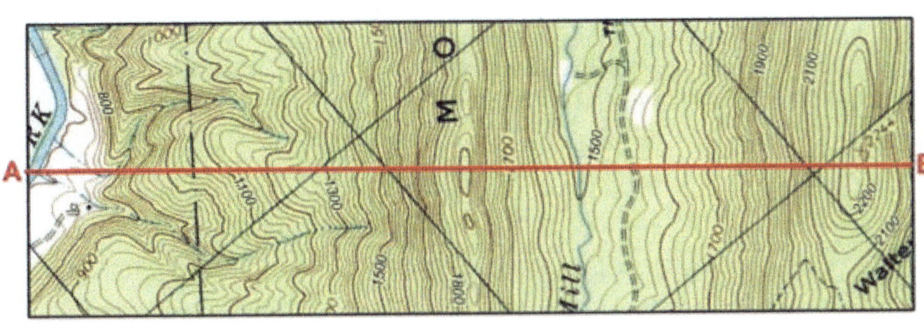

1:24,000 (One inch = 2,000 feet)
Contour interval 20 feet

0 1,000 2,000 3,000 4,000 Feet

Chapter 3

Thinking About Earth

Back when I was a middle school student, which was in the 1970s, dinosaurs were often thought of as slow, stupid, overgrown, cold-blooded lizards. Dinosaur displays in museums made dinosaurs look rather awkward and cumbersome. Since that time, our thoughts about dinosaurs have radically changed. We now know that dinosaurs were much more mobile than we once thought, that they may have been warm-blooded, and that many species lived in social groups and raised their young rather than just abandoning their eggs. It appears that, in many ways, dinosaurs were more similar to today's birds than to today's lizards. How have we learned such things? To answer questions like this, we must give more thought to how science works in Earth sciences.

Objectives

After studying this chapter and completing the exercises, you should be able to do each of the following tasks, using supporting terms and principles as necessary.

1. Describe the Cycle of Scientific Enterprise by describing each step in the cycle, and explaining how the steps relate to each other.
2. Distinguish between theories and hypotheses, giving examples from Earth sciences.
3. Distinguish between experimental science and historical science.
4. Explain why scalability, accessibility, and complexity are challenges for much of the work done by Earth scientists.
5. Express ways in which observing the creation might move a Christian to worship God.
6. Identify ways in which Earth scientists can love God and love people through their work.
7. Describe three ways in which God has revealed himself to us.
8. Describe how Christians can respond when there appears to be a conflict between Christianity and science.
9. Explain why good stewardship of Earth is a biblical concept.
10. Distinguish between renewable and nonrenewable natural resources.
11. Explain what is meant by sustainability.

Vocabulary Terms

You should be able to define or describe each of these terms in a complete sentence or paragraph.

1. controlled experiment
2. Cycle of Scientific Enterprise
3. experiment
4. experimental science
5. greenhouse effect
6. historical science
7. hypothesis
8. natural resource
9. nonrenewable natural resource
10. plate tectonics
11. oceanic circulation
12. renewable natural resource
13. scalability
14. scientific fact
15. stewardship
16. theory

3.1 The Cycle of Scientific Enterprise in Earth Sciences

The term *science* has been defined in many different ways. A basic definition would be that science is our attempt to understand the physical universe, using observation, experiments, and reason. Like scientists in other fields, such as physics or chemistry, Earth scientists use observation, experiments, and reason in order to come to a better understanding of how our planet works.

There are many ways of doing science, but a good overview of how science works is given by the *Cycle of Scientific Enterprise*, which is diagrammed in Figure 3.1. The Cycle of Scientific Enterprise is the process by which scientists form theories and hypotheses, test those hypotheses with experiments, and analyze the results.

We can get a better idea of how science works in Earth sciences by going around the cycle step by step. We can start on the left side of the diagram with the concept of scientific facts.

3.1.1 Scientific Facts

A *scientific fact* is a scientific statement that is supported by a large amount of evidence and that is correct as far as we know. Some examples of facts in Earth sciences include:

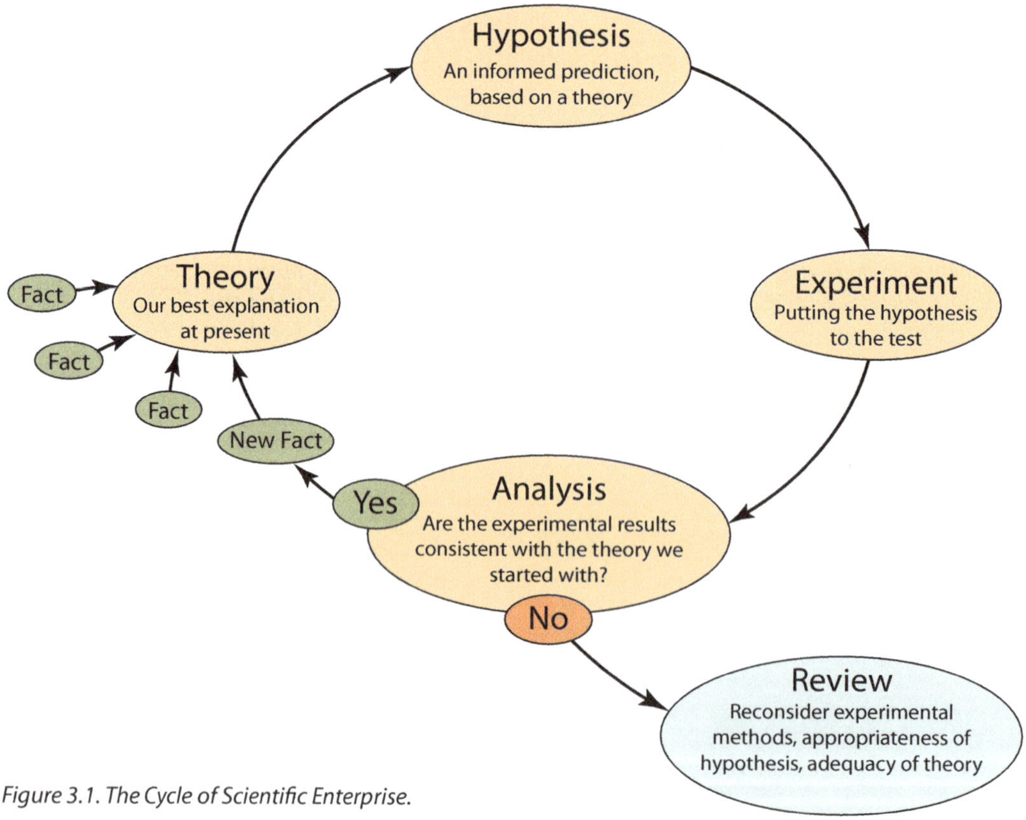

Figure 3.1. The Cycle of Scientific Enterprise.

- Seawater contains, on average, 3.5% dissolved salts, primarily sodium chloride (NaCl), commonly known as table salt.
- Earth's atmosphere is composed primarily of nitrogen, oxygen, and argon, with a variable amount of water vapor, and minor amounts of other gases.
- The temperature inside Earth increases with the depth below the surface at a rate of about 25°C per kilometer of depth (about 72°F per mile). Therefore, the temperature at a depth of ten kilometers (about six miles) is 250°C warmer than the average temperature on Earth's surface.

3.1.2 Scientific Theories

However, science is much more than the mere collection of facts about the universe. We cannot really understand the world God has made just by saying things like "seawater is salty," or even by measuring the precise amount of sodium chloride that is dissolved in a liter of seawater. A scientific *theory* is an explanation that seeks to account for all related facts. A theory is a mental model that explains how something in the universe works. Sometimes theories are stated with mathematical equations, such as Newton's Second Law of Motion. This equation tells us that the acceleration of an object is directly proportional to the force acting on that object, and inversely proportional to the object's mass:

$$a = \frac{F}{m}$$

In other cases, theories are more explanatory, such as theories about what causes volcanoes to erupt or why certain fossils occur in some types of rocks but not in others (though explanatory theories often involve some mathematics as well).

Some examples of important scientific theories in Earth sciences are:

- The theory of *plate tectonics*, which describes Earth's lithosphere as a mosaic of continent-sized plates that move around at rates of a few centimeters per year, and explains why these movements occur. Some of the plates are moving away from each other (North America is moving away from Europe), some are colliding (India is slowly moving northward into Asia, as seen in Figure 3.2), and others

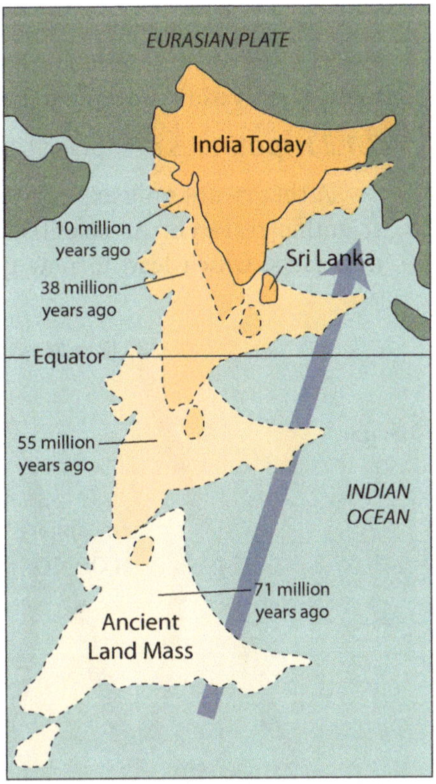

Figure 3.2. According to the theory of plate tectonics, the lithosphere beneath both continents and seafloor moves around on Earth. One result of this is the collision of India into Asia, which has led to the uplift of the Himalayas.

slide past each other. The theory relates the plate movements to other phenomena such as earthquakes and volcanoes.

- The *greenhouse effect* theory tells us that certain gases in Earth's atmosphere, such as water vapor and carbon dioxide (CO_2), absorb electromagnetic radiation in a way that keeps Earth's surface considerably warmer than it would be if Earth's atmosphere did not contain those gases. Without the greenhouse effect, Earth would be a frozen ice world. The theory explaining the greenhouse effect is an area of active research. Figure 3.3 shows a basic diagram illustrating how the greenhouse effect works.
- The theory of *oceanic circulation* explains how Earth's rotation, wind, seawater density, and gravity interact to drive ocean currents. The fact has been known since ancient times that there are large-scale currents in the waters of the oceans—massive rivers of water within the seas. These currents move at speeds of a few miles per hour. The first map of the Gulf Steam, shown in Figure 3.4, was compiled by Benjamin Franklin.

Each of these theories not only describes some natural phenomenon, but also *attempts to explain why* some aspect of the natural world behaves as it does. These theories are all well tested, are accepted by most scientists, and seem to explain many observations about Earth.

3.1.3 Hypotheses and Scientific Research (Experiments)

A theory not only explains certain facts, but provides a foundation for further scientific research. As scientists investigate the world, they continually uncover new facts, which lead to new questions about how those facts relate to existing theories. A *hypothesis* is a proposed answer to a scientific question, often stated in the form of a prediction of what a scientist expects to find next. The plural of *hypothesis* is *hypotheses*.

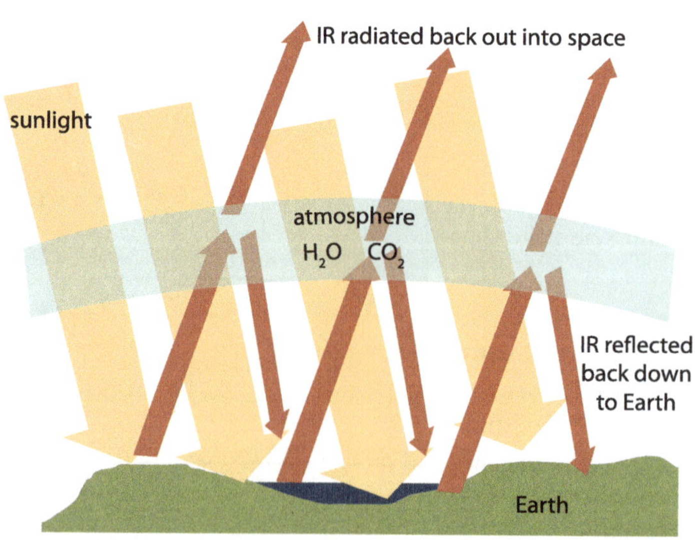

Figure 3.3. The greenhouse effect works because water vapor and carbon dioxide in the atmosphere absorb infrared radiation emitted by the warm Earth, preventing some of that energy from leaving the planet.

A hypothesis is not a wild guess. It is a prediction that is based on everything the scientist currently knows about the topic at hand. For example, based on our present understanding of the greenhouse effect, most scientists who study Earth's climate predict that as the amount of carbon dioxide increases in Earth's atmosphere—primarily as a byproduct of the burning of fuels such as gasoline and coal—the greenhouse effect will become more intense, and our climate will become warmer. This is a prediction based on our current theory of the greenhouse effect. We will take a closer look at the complexities of the greenhouse effect and potential climate change in a later chapter.

Figure 3.4. Benjamin Franklin's 1770 map of the Gulf Stream, an ocean current in the North Atlantic Ocean. The Gulf Stream flows to the northeast off the east coast of what is now the United States.

A hypothesis is a statement that can be tested so a scientist can confirm or disconfirm a particular hypothesis. Often the test of the hypothesis is in the form of an experiment. A *scientific experiment* is an organized process by which a scientist tests a hypothesis. Often an experiment is conducted in a laboratory, where the scientist can attempt to establish ideal conditions for the study of some natural phenomenon. For example, Figure 3.5 shows a tornado created in a laboratory. This laboratory tornado is in reality only a model of tornadoes that exist in nature; it is much smaller than natural tornadoes, with lower wind speeds. But it still enables scientists to study wind currents within the spinning column of air, and the damage tornadoes can cause to buildings and other structures.

Figure 3.5. A laboratory simulation of a tornado.

3.1.4 Analysis

After an experiment is conducted, it is important for scientists to analyze the results to see whether the initial hypothesis was confirmed. Confirmation of the hypothesis adds support to the theory the hypothesis was derived from. A successful theory will be supported by the results of a large number of experiments. A theory that does not lead to successful predictions is considered weak, and will eventually need to be revised or discarded.

Sometimes the results of an experiment are not what the scientist expected. This does not automatically mean that the hypothesis—or the theory it came from—were incorrect. Most theories account for most observations and experimental results, but not all. It could be that the results of the experiment will result in the modification of the theory. In this case, the unexpected results could lead to additional experiments that will end up strengthening the theory. Another possibility is that the experiment was somehow faulty. It could be that the experiment was poorly designed or that the equipment used in the experiment was defective. As scientists review the results of their experiments, they might discover and correct these flaws, and then conduct the experiment again.

Two of the most important activities done by scientists are actually reading and writing. You may have a mental picture of scientists in a laboratory filled with equipment or out in nature making observations, but scientists actually spend a large amount of their time reading technical papers in scientific journals, which are magazines with articles that are critiqued by other scientists before they are published. Reading these articles and interacting with other scientists allow scientists to keep up with the latest research done by other scientists from all over the world. Most scientists also spend a lot of time writing, whether writing reports that will be used inside of their own research group, or writing papers about the results of their own research that will be published in the scientific journals. The publication of articles in scientific journals allows other scientists to review and critique the results of scientific studies.

Learning Check 3.1

1. Draw a sketch of the Cycle of Scientific Enterprise, with arrows and key words.
2. Give examples of theories in geology, meteorology, and oceanography. Try to use theories that were not mentioned in this section. (Hint: you can page through this book.)
3. Explain why reading and writing are important aspects of doing science.

3.2 Experimental Science and Historical Science

Work in Earth sciences often proceeds as described in the previous section, as scientists conduct experiments to determine whether a hypothesis is valid. These experiments may be performed in a laboratory, as we saw in Figure 3.5, or may be done outdoors in a natural setting.

An example of an experiment conducted in the field, rather than in a laboratory, is an experiment

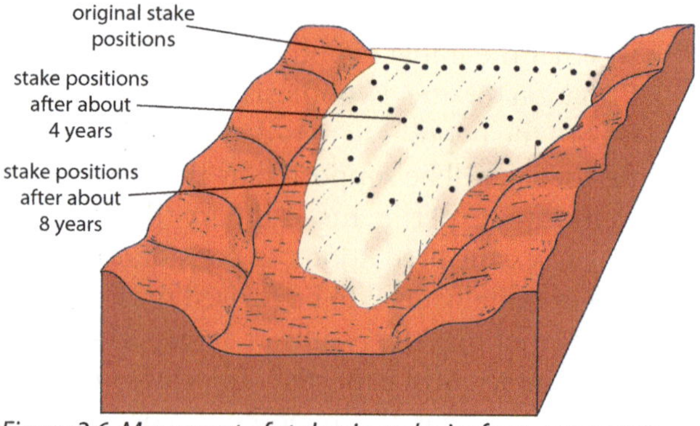

Figure 3.6. Movement of stakes in a glacier from one year to the next.

conducted in the Swiss Alps in the late 1830s through the early 1840s by a geologist named Louis Agassiz (pronounced LOO-ee AG-uh-see). The higher parts of the Alps have areas of snow year-round. Based on observations of a hut that had been built on the surface of a glacier in the 1820s, Agassiz formed the hypothesis that the ice in the glacier actually moved downhill, but at a rate too slow to be observed directly. Agassiz placed a series of wooden stakes in a straight line across the ice, and measured how the stakes moved over time. He demonstrated that the ice moved downhill at a rate of a few centimeters per day, which works out to over a hundred meters per year. But he also discovered something that went beyond his initial "glacier ice moves" hypothesis. Stakes pounded into the ice near the edges of the glacier move more slowly than stakes in the center of the glacier, as illustrated in Figure 3.6. One reason for this is friction between the moving ice and the rock walls of the glacial valley; the ice in the center of the glacier is not affected as much by friction along the edges of the glacier.

Agassiz's experiment is an example of *experimental science.* When doing experimental science, scientists test hypotheses by conducting planned, well-controlled experiments. Experimental science is valuable for evaluating many processes that are occurring today, but is not as directly applicable to events that occurred in the past. It is not possible to recreate historical events, such as the formation of the solar system, the deposition of the rock layers that cover much of Earth's continents, or the climate changes that led up to the ice ages. This

Figure 3.7. When archeologists reconstruct historical events based on the evidence they find in an excavation, they are practicing historical science.

does not mean that these events cannot be studied scientifically, and it does not mean that scientists cannot form hypotheses and try to confirm them. But it does mean the investigation of historical events requires the use of different methods.

Geologists often study Earth using appropriate to what we call *historical science*. In sciences such as chemistry and physics, most hypotheses are tested by scientific experiments. In geology, and to some extent in climatology and oceanography, as well as in sciences such as astronomy and archeology, the "experiment" has already been run in nature. In these cases, the job of the scientist is very much like the job of a detective in solving a crime. A detective looks for evidence at a crime scene, and then attempts to reconstruct the events that happened. Likewise, the scientist's goal, when doing historical science, is to consider all of the evidence and then to put together a credible story of what happened in the past. As scientists do further investigations, additional evidence might be uncovered that adds to the story, or that perhaps contradicts the story.

Archeologists study events from human history by carefully excavating sites where humans once lived or worked. A photograph of the site of an archeological excavation is shown in Figure 3.7. Often there is no written record telling the history of the people who lived there, so the only way to reconstruct what their lives were like is to examine various things they left behind—such as burial sites, fireplaces, pottery fragments, and the remains of their houses—and use these to figure out how the people lived and what happened to them. The archeologist then tries to put together a compelling story about the lives and history of the people who dwelt at the site. This story is a scientific hypothesis that can be tested by doing additional excavation or by comparing the results of the excavation to archeological facts from other nearby sites. This additional research may add details to the archeologist's story, or even lead to a complete rewriting of the story.

Archeologists study artifacts left by human beings. Going back further in time, many of the events studied by geologists occurred before there were any humans on Earth, but geologists use the same sort of historical reasoning used by archeologists. The Grand Canyon in Arizona contains many layers of rock, as shown in Figure 3.8. Each of these layers has properties that are distinctive, and geologists use observations and reason to infer how each layer formed. Some of the layers appear to have been formed in warm, shallow

Figure 3.8. Geologists use historical scientific reasoning to interpret the history of Earth, such as the events that led to the formation of the Grand Canyon.

Figure 3.9. The Coconino Sandstone in the Grand Canyon has numerous features that indicate it was formed in a desert rather than under water.

seas with an abundance of living organisms. Others formed in deeper water, in fresh water, or along coastlines. Geologists use multiple types of evidence to make these interpretations, including analysis of the types of fossil organisms that are present in each layer.

Let's take a closer look at one of the layers of the Grand Canyon to get a better picture of how scientists use historical scientific reasoning. Geologists have hypothesized that a rock layer called the Coconino Sandstone near the top of the Grand Canyon was formed by sand dunes in an ancient desert. This layer is shown in Figure 3.9. The sand-dune hypothesis was based on comparisons of the solid rocks of the Coconino Sandstone to sandy deserts, such as sand dune areas found in parts of the Sahara Desert. In order to test the sand-dune hypothesis, geologists looked for the sorts of things one would find in a desert—such as tracks of lizards, scorpions, spiders, and millipedes preserved in the rocks—and found these tracks in abundance. These footprints could not have formed under water, such as in a stream or in an ocean. On the other hand, they have not found any compelling evidence that would indicate this layer was formed under water. For instance, the rocks of the Coconino Sandstone do not contain any fossils of water-dwelling organisms, such as seashells or fish.

Both experimental science and historical science are used to come to a better understanding of the world God has given us as a home. Both methods fit within the Cycle of Scientific Enterprise, which is our method of taking scientific facts and developing theories through a process of forming hypotheses, testing of hypotheses, and analysis of results. Both approaches to science give us tentative answers that lead to additional hypotheses that can be tested and evaluated.

Learning Check 3.2

1. Explain why some scientific hypotheses cannot be tested by conducting an experiment.
2. In what ways is historical science like detective work?

3.3 Challenges in Studying Earth

Earth scientists face additional challenges as they seek to understand Earth, past and present. The first of these challenges is that of *scalability*. Scalability in Earth sciences is the ability of processes in a scale model to work in the same way as processes do in nature. Earth scientists face this problem when they try to create models of phenomena that occur on Earth, such as the tornado model shown in Figure 3.5. The tornado in the laboratory will behave like a real tornado in some ways, but not in others.

The indoor model of San Francisco Bay shown in Figure 3.10 is about 400 feet long. The hydrologists (water scientists) and engineers who created this model had to include numerous design features to ensure that the little waves and currents in this gigantic model behaved similarly to waves and currents in the real San Francisco Bay. They did a good job, but the model has its limitations.

A second challenge is that of *accessibility*, which is the ease or difficulty of getting to a place. Often Earth scientists have to study some place or process that they cannot directly observe. For example, it is impossible at present for geologists to send instruments into the core of Earth. Another example is that meteorologists

Figure 3.10. An indoor model of San Francisco Bay.

Figure 3.11. The lava lake in the crater of Mount Nyiragongo, Democratic Republic of Congo, is an inaccessible place. The red areas are molten rock, and the black areas are thin crusts of crystallized rock on top of the lava lake.

are able to observe tornadoes from the outside, but cannot get inside a tornado to make direct measurements. Despite these challenges, we do know quite a bit about the core of Earth and the interior of tornadoes.

There are other places on Earth that are accessible, but not easily so. The deepest parts of the ocean are over ten kilometers (six miles) deep, and only a handful of people have briefly descended to those depths. There are also places on Earth's surface that are difficult to access, such as the bottoms of rivers, places far from civilization, and the lava lakes that exist in the craters of some volcanoes, such as the one shown in Figure 3.11.

A third challenge is *complexity*. In a laboratory experiment, scientists try to simplify conditions so they can focus on just one thing, and see how that one thing changes as they change one other variable. That is what is meant by a *controlled experiment*. However, in nature conditions are almost never simple. Complexity is the opposite of simplicity; it is the state of having many different things happening at one place or one time. An example of a complex system is the greenhouse effect mentioned earlier. Figure 3.3 shows a simplified overview of the greenhouse effect, caused by the presence of carbon dioxide and water vapor. But carbon dioxide and water vapor are not the only greenhouse gases in Earth's atmosphere. Methane (CH_4) and ozone (O_3) are also important greenhouse gases, even though there are only tiny amounts of these gases present. The influence of these gases on the overall temperature of Earth's atmosphere is not completely understood.

An additional complexity in Earth's climate is the interaction between the greenhouse effect and clouds in the atmosphere. If an increase in the amount of greenhouse gases in the atmosphere causes Earth's temperature to increase, that will lead to greater evaporation of water from the oceans. More evaporation from the oceans will lead to greater amounts of cloud cover. More clouds forming will lead to more sunlight being reflected out to space, because clouds are more efficient at reflecting light than either the oceans or land surface. More sunlight being reflected out to space means Earth will cool. So which is the more important factor in determining what will happen to Earth's temperature in the future: the increased heating due to an increase in greenhouse gases, or cooling due to an increase in cloud cover? Climate scientists make mathematical models of Earth's atmosphere on powerful computers to try to answer questions such as this, and most of them believe that the temperature of Earth's atmosphere will increase in upcoming years.

It is amazing to live in a world that is governed by natural laws, so that we can understand it, but is at the same time so incredibly complex that we cannot understand it fully. The challenges outlined in this section make doing Earth science difficult at times. For many Earth scientists, however, these challenges are part of what makes their work so interesting.

Learning Check 3.3

1. Explain how scalability, accessibility, and complexity make Earth science work more challenging.
2. Describe a place on or in Earth that seems inaccessible, and suggest a way scientists might be able to study that place.

3.4 Christianity and Earth Sciences

Study of the natural world should move us to worship its Creator. Psalm 19:1 says, "The heavens declare the glory of God and the sky above proclaims his handiwork." It is not just the stars and galaxies that bring glory to God, but things on Earth such as floods, thunderstorms, waves rolling on beaches, forests, and volcanoes. I, like many Christians, have been moved to sing hymns of praise to God while being astonished by his creation:

O Lord my God, when I in awesome wonder
Consider all the worlds thy hands have made,
I see the stars, I hear the rolling thunder,
Thy power throughout the universe displayed:
Then sings my soul, my Savior God, to thee:
How great thou art! How great thou art![1]

1 From the hymn "How Great Thou Art" by Carl Gustav Boberg, translated from Swedish by Stuart K. Hine.

3.4.1. Doing Science is a Way to Love God and Love People

In all types of work, Christians are called to obey the two greatest commandments: "You shall love the Lord your God with all your heart and with all your soul and with all your mind," and "You shall love your neighbor as yourself" (Matthew 22:37, 39). The vocations of meteorologist, oceanographer, and geologist are all ways to love God and love people. We love God in our work by doing it with integrity and diligence, being faithful to walk with Christ wherever God places us. Our work is also one of the primary ways in which we love and serve people, and Earth scientists serve people by making weather forecasts, finding energy and mineral deposits, conserving fishing resources in the oceans, finding and protecting drinking water resources, and in many other ways.

3.4.2. Science and Truth

God reveals truth to us in multiple ways. As Christians, we believe that the supreme way in which God has revealed truth to us is in Jesus Christ. Jesus taught his disciples, "I am the way, and the truth, and the life," and "Whoever has seen me has seen the Father" (John 14:6, 9). When we look at Jesus, we see what God is like in ways we could see in no other way.

God has also revealed himself in his word, the Bible, and in the creation. There are many things that we would never know if God had not told us. We might be able to deduce from nature that there must be a God, but we would never know that Jesus died on the cross for our sins unless God had revealed it to us in the Scriptures.

The idea that God has revealed himself in the creation is repeated in the New Testament, in Romans 1:20: "For his invisible attributes, namely, his eternal power and divine nature, have been clearly perceived, ever since the creation of the world, in the things that have been made."

In what ways has God's "eternal power and divine nature" been revealed in "the things that have been made"? At the most basic level, the very existence of the universe speaks of the existence of a Creator. If we ask questions such as, "Why is there a universe?" or "How did the universe come into being?", there are only a few possible answers:

- One answer is that the universe has always existed, but this doesn't really explain *why* it is here. Also, there is very strong evidence in astronomy that points to the universe having a beginning, perhaps around 13.8 billion years ago, so there are few scientists who believe that the universe has existed forever.
- Another answer is that the universe was created by a larger universe outside of the universe we can observe. Sometimes this bigger universe is called the multiverse. But this only pushes the issue of why there is a universe back one step by raising the new question, "Why is there a multiverse?"
- A third answer is the answer given by Christianity and some other religions. This answer is that there is something outside of the universe that is not part of a

larger universe. We call that something "God," and in this third answer, it is God who has created the universe. Some people then ask, "What created God?", but I don't think that is a very good question because it implies that God is part of the creation. Christians have traditionally believed that God is not part of the creation, so this third answer is quite different from the second answer.

For many scientists, as well as many other people, this sort of reasoning provides strong support for the belief in the existence of God. Certainly belief in God is not unscientific.

3.4.3. Do Christianity and Science Conflict?

I really hope you remember a lot of things from taking this course in Earth science, but if you only remember one thing from this book, I hope it is this:

All truth is God's truth.

When we study God's word, we are studying God's truth. When we study science, we are also studying God's truth.

Our problem, however, is that we are humans, and that means that we are limited in our capabilities to understand either God or his creation fully. This limitation comes because we are finite—only God has infinite knowledge—and because we are tainted by sin.

It should be clear that we sometimes misunderstand nature. People used to believe that frogs arise spontaneously from mud; now we know that frogs come from frog eggs, which are laid by female frogs. In other words, life comes from life. I could give many examples such as this, but the point is that we have had, and almost certainly still do have, scientific theories that are incorrect.

Both nature and the Bible are subject to human interpretation. But that does not mean that all interpretations are equally valid; there are good interpretations and there are poor interpretations. An example of a poor interpretation of Scripture that has now been almost universally abandoned is the use of verses such as Psalm 93:1 to support the idea that Earth could not possibly orbit the sun: "The world is established; it shall never be moved."

To medieval theologians, if the world "shall never be moved" then it certainly could never orbit the sun. The sun, therefore, must orbit Earth. Today, it is understood that Psalm 93 is poetry, and not everything is meant to be taken literally. If we were to take everything in Psalm 93 literally, we would have to believe that God is wearing a robe with a belt and that water has a voice. Once Bible scholars took a closer look at Psalm 93 as a whole, the perception that there was a conflict between the Bible ("it shall never be moved") and science ("Earth moves") faded away.

There are, however, still a number of people, both Christians and non-Christians, who believe that there continues to be a conflict between science and Christianity. Because of this perceived conflict, there are a number of non-Christians who believe that science has somehow proved the Bible wrong. One area of perceived conflict is about the age of Earth. Before the modern age, most Christians assumed that the Bible taught that Earth had been created by God roughly around 4,000

BC. Beginning in the 1700s, many geologists began concluding that Earth must be considerably older than that, and by the early 1800s most geologists believed that the age of Earth was many millions of years old. The generally accepted value for Earth's age at present is about 4.6 billion years.

So which is true?

To answer this question, we need to go back to the statement that *all truth is God's truth*. If there is a conflict between how some people interpret Genesis (Earth is 6,000 years old) and how scientists interpret Earth history (Earth is 4.6 billion years old), then either we don't understand nature correctly, or we don't understand the Bible correctly. For now, we need to approach these "conflicts" with grace and humility.

Those Christians who continue to hold to the idea that Earth is only about 6,000 years old understand Genesis as teaching that Earth is young. Along with this, there are Christians in the sciences who believe that there is scientific evidence that indicates that Earth is young rather than billions of years old. I respect these people for their commitment to the Bible, their love for God, and their intelligence, but I also disagree with them.

The issue of the age of Earth can be quite divisive between Christians. Here is the perspective that will be taught in this textbook: the Bible does not clearly teach how old Earth is, and scientific evidence strongly points to an Earth that is billions of years old. Most Christian geologists agree with this perspective on the scientific evidence, and there are many Bible scholars who agree that the Bible does not require us to believe Earth is young. It is sort of like the situation in the 16th century with Bible verses that seemed to prohibit Earth being in motion around the sun. When Bible scholars took a closer look at the verses, it became clear that the Bible does not teach that Earth is motionless. In the same way, I believe a closer look at Genesis reveals that the Bible does not teach that Earth is young.

Whatever you think about the age of Earth, remember that *all truth is God's truth*. There may seem to be conflicts at times between the Bible and science, but the level of conflict is not nearly as great as some people think. In the end, when we understand everything, the conflicts will disappear.

Learning Check 3.4

1. Find a psalm that declares God's glory in the creation and write a selection of verses from that psalm. Your psalm should be one that was not discussed in this section.
2. How can an Earth scientist's work demonstrate love for God and love for people?
3. Name three ways in which God has revealed truth to us.
4. Explain what is meant by the statement, "all truth is God's truth."

3.5 Stewardship of Earth

When God gives us something of value, we are responsible for using that thing with wisdom. *Stewardship* is the wise and prudent management of the things entrusted to us. We are all entrusted with various things, such as our time, money, talents, and health. Some people are good stewards of what God brings into their lives, and in general this works out for their good. Others are poor stewards of their time, money, talents, or health, and this usually works out badly not only for those who are poor stewards, but for those who depend on them.

3.5.1 Earth Stewardship in the Bible

The first area of stewardship mentioned in the Bible is the stewardship of Earth itself. According to Genesis 1 and 2, God not only created our world "very good," but placed humans in a special stewardship role over the creation. The foundation for this idea is in Genesis 1:28, which states God's intention for the newly created human pair: "And God blessed them. And God said to them, 'Be fruitful and multiply and fill Earth and subdue it and have dominion over the fish of the sea and over the birds of the heavens and over every living thing that moves on Earth.' "

Most Bible scholars believe this command to subdue and have dominion over Earth did not apply just to Adam and Eve, but to all of their descendants, including us. Some people have read this verse, with the words "subdue" and "dominion," and concluded that the Bible teaches that humanity can do whatever they want to Earth. But the picture presented in Genesis is one in which Adam and Eve are *vice-regents*, that is, people who rule in someone else's place. God is the king of the universe, but he placed Adam and Eve—and their descendants—as his representatives over the creation. This means that humans are to rule over the creation as God would; to be stewards of creation for God's glory, for the good of all the people of Earth, and for the flourishing of the creation itself. This points once again to the two greatest commandments in Scripture, to love the Lord our God with all our being, and to love our neighbors as ourselves.

The account of the Garden of Eden in Genesis 2 gives us an additional perspective on what it means to have dominion over Earth. Genesis 2:15 says, "The Lord God took the man and put him in the garden of Eden to work it and keep it." A gardener may do things to his garden to improve its ability to yield a crop, such as tilling, irrigating, or adding fertilizer. The gardener does these tasks in order to make the garden more fruitful. We would consider a gardener to be a poor steward of his garden if he kept it in such a way that the soil eroded away, the water became polluted, and weeds crowded out the desired plants.

It is hard to read the first two chapters of Genesis without getting a sense of a world of abundance where life flourishes. For instance, Genesis 1:20–22 describes the creation of sea life and flying things:

> *And God said, 'Let the waters swarm with swarms of living creatures, and let birds fly above Earth across the expanse of the heavens.' So God created the great sea creatures and every living creature that moves, with which the*

waters swarm, according to their kinds, and every winged bird according to its kind. And God saw that it was good. And God blessed them, saying, 'Be fruitful and multiply and fill the waters in the seas, and let birds multiply on Earth.'

The statement that the creation was to "swarm with swarms" of all sorts of living creatures tells us that God did not create Earth to be sparsely populated, but to have great numbers of many different types of organisms in many different places.

3.5.2 Earth Resources

The world God has given us is rich in living and nonliving *natural resources.* A natural resource is something provided by Earth that human beings can use. Living resources include the plants and animals we use for food, the cotton we use for clothing, and the trees we use for lumber. Nonliving resources include water, air, soil, and ores that are mined to obtain metals.

An important distinction that needs to be made in regard to natural resources is the difference between *renewable* and *nonrenewable* resources. A renewable natural resource is one that can be replaced as it is used. Trees are an example of a renewable resource. We can harvest trees to use as lumber, pulp for making paper, or fuel, but new trees can grow to replace the ones that were harvested. Likewise, water in streams and most water in groundwater supplies are renewable resources. We can use water in a stream knowing that the supply will be replenished by future rainfall or snow.

This does not mean that renewable resources are available to us in an unlimited way. The water in the Colorado River in the arid southwestern United States is renewable. Every winter, it snows in the Rocky Mountains, and every spring and summer, that snow melts and some of that meltwater ends up in the Colorado River. But in most years the demands for water—primarily for irrigation and drinking water—are greater than the amount of water flowing in the river, and the result is that the Colorado River is usually completely dry before it reaches its outlet in the Gulf of California in Mexico.

A nonrenewable natural resource, on the other hand, is one that is not replaced when it is removed from nature for human use. When a mineral, such as iron ore, is mined and removed from Earth, that iron ore is not immediately replaced in Earth's crust by any geological process.

One example of a nonrenewable resource that people use a lot of is coal, which is burned as a fuel in power plants to produce electricity. Many billions of tons of coal are mined from Earth every year, but there are only so many billions of tons of coal in Earth's crust. The United States has more coal than any other country, but at the rate we are mining it, is will all be consumed within a few hundred years. That sounds like a long time from now, but in discussions about energy and resources, we need to think not only about our present needs, but about how our present actions will affect future generations.

3.5.3 Sustainability

Good gardeners, good farmers, and good ranchers will take care of their land in such a way that it will remain productive for a long time. *Sustainability* is the ability to maintain the productivity of the creation for a long, or even indefinite, amount of time. It is the ability to use natural resources to meet the needs of today without reducing the ability of future generations to have their needs met as well.

We can apply the principle of sustainability to our own lives. An example of this would be an individual taking care of her health in a sustainable way. If a person has a good diet, a good sleep routine, and exercises, she is likely to be able to sustain good health over much of her lifespan. If, on the other hand, a person has a long-term lifestyle of getting only three hours of sleep per night, has a diet of French fries and donuts, and doesn't exercise, her health will deteriorate. The long-term results of this lifestyle will likely be poor health and an early death. In other words, she will be living unsustainably.

The principle of sustainability works the same way in regards to Earth resources, both living and non-living. A sad example of people exploiting a natural resource in an unsustainable way is the overfishing of cod that has occurred in the northwest Atlantic Ocean. Fishermen had been fishing for cod in the Grand Banks area of the North Atlantic, off the coast of eastern Canada, since the 1500s, and had been doing so in such a way that the population of cod never became depleted. There were always enough fish left over to reproduce and make up for those that were being harvested; in other words, fish were being harvested in a sustainable way. Technological improvements, such as larger fishing trawlers and improved,

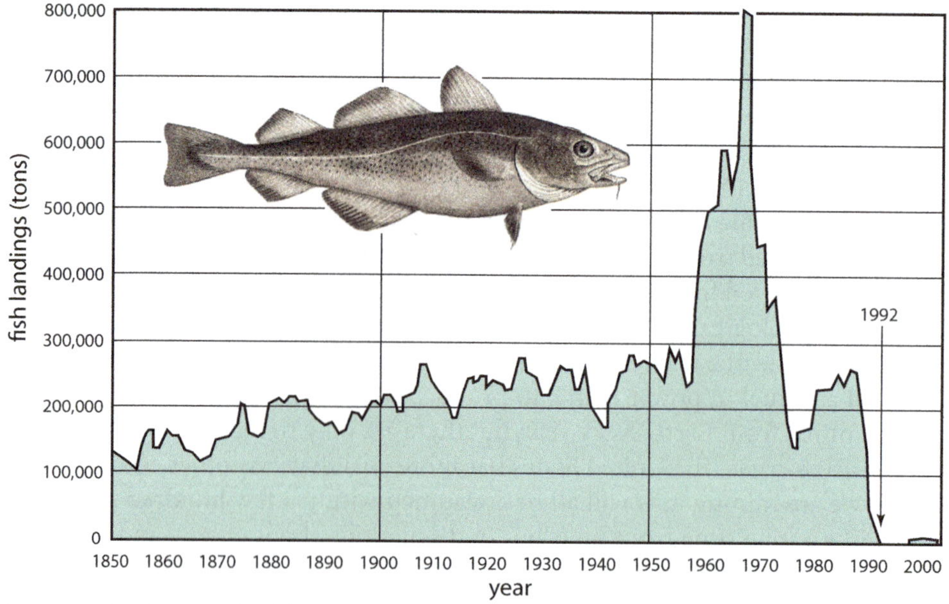

Figure 3.12. The harvest of cod from the northwest Atlantic Ocean was sustainable until sometime in the 1950s, and then climbed dramatically to unsustainable levels in the 1960s. This led to a collapse of the fishery to less than 1% of historic levels beginning around 1990.

larger nets, were introduced in the 1950s, and for a few years this led to significantly greater harvests of fish. On the graph on Figure 3.12, you can see that the number of tons of cod caught off the east coast of Canada rose dramatically starting in the late 1950s, only to collapse to around 1% of earlier catches by the early 1990s.

There were a number of factors that led to unsustainable exploitation of the northwest Atlantic fisheries. Not only did improved fishing technology allow for overharvesting of cod, but people did little to change their fishing practices even when it became apparent that the cod population levels were plummeting. In 1992, the Canadian government banned fishing for cod in its waters. People expected the cod population to recover quickly, but that did not happen. Somehow, overfishing not only eliminated most of the cod, but also changed the ecology of the ocean in such a way that cod could no longer flourish. Today, the population of cod in these waters is still well below historical levels.

3.5.4 The Degradation of Creation

Earth was designed in such a way that it exhibits astounding resilience. There are natural systems that regulate temperatures, purify water, move nutrients around the planet, and renew topsoil needed for plant growth. When things are working properly, the living world—including humans—can prosper, as God intended.

Humans are capable of degrading the creation in ways that greatly diminish its ability to support plants, animals, and even us. Let's take a look at ways that these degradations affect the four Earth systems described in Section 1.1.

- *Degradation of water*—Water can be polluted by many things, such as human waste (sewage), toxic waste from industrial sites, and pesticides used in agriculture. For a long period of time from the late 1800s until the 1970s, the Cuyahoga River in Ohio was so polluted that the river would occasionally catch on fire. These disasters helped to raise citizen awareness about water pollution, which led to government regulations such as The Clean Water Act in the United States. The stream in Figure 3.13 has been polluted by iron associated with a nearby coal mine, which has made the water unsuitable for fish, wildlife, agriculture, or human consumption.
- *Degradation of air*—Air can become polluted by both visible and invisible pollutants. These pollutants come from a

Figure 3.13. A polluted stream near a coal mine.

variety of sources, but the primary sources are the burning of fossil fuels, such as coal in power plants and gasoline in automobiles. The Great London Smog of 1952 was caused by the burning of coal in many thousands of stoves, factories, and power plants. The smog was so thick that visibility was reduced to just a few feet. Some researchers estimate that as many as 12,000 people died in London because of that event. The word "smog" is a combination of "smoke" and "fog." Figure 3.14 shows a coal-fired power plant, which produces electricity.

Figure 3.14. A coal power plant in Poland. Most of the visible emissions are just water vapor, but invisible gases include sulfur dioxide (SO_2) and various nitrogen oxides. Combustion of coal also produces a tremendous amount of carbon dioxide, a greenhouse gas.

- *Degradation of life*—Genesis 1 paints a picture of a planet teeming with life; on the land, and in the air and sea. Largely because of human activities, however, thousands of species have been driven to extinction, and many more types of organisms could share the same fate in the upcoming century. If humans were managing Earth as intended, our activities would tend to promote the fruitfulness and rich diversity of the living world rather than its destruction.
- *Degradation of soil*—Healthy soil is essential for agriculture because plants depend on the soil for water, nutrients, and as a place to be anchored to the ground. In many agricultural areas around the world, soil is eroding away due to poor agricultural practices, as shown Figure 3.15.

Figure 3.15. Soil erosion in a wheat field in eastern Washington. The topsoil is essential for abundant plant growth, but is the first part of the soil to be washed away when the soil is bare.

Other degradations include reduction of habitat for a diverse range of plant and animal species, human-caused expansion of deserts, and accumulation of wastes in landfills and other places.

The degradation of the creation runs counter to the picture in Genesis of a flourishing, abundant world. It is inevitable that as humans use natural resources, the world will be affected, but human sin has meant that much of our impact on God's world is negative. God's design is for humans to have a positive impact on Earth, protecting and stewarding the rest of the creation and living in harmony with it.

Learning Check 3.5

1. Explain ways in which stewardship of Earth is like stewardship of something in your life, such as money or time.
2. Explain what is meant in Genesis 1:28, when Adam and Eve were given the command to have dominion over Earth.
3. Using examples, explain the difference between renewable and nonrenewable natural resources.
4. Why is it important that humans use natural resources in a sustainable way?

Chapter 3 Exercises

Answer each of the questions below as completely as you can. Write your responses in complete sentences unless instructed otherwise.

1. Using the concept of the Cycle of Scientific Enterprise, describe how Earth scientists study Earth.
2. How do Earth scientists use experiments to study Earth?
3. Why is it sometimes necessary for Earth scientists to use historical science rather than experimental science?
4. Read Job 38–39 and list eight ways in which God is glorified in the creation.
5. Suggest ways in which an Earth scientist's work can serve people. Be as specific as possible, and don't use examples from this chapter.
6. When learning about Earth sciences, why is it important to remember that "all truth is God's truth?"
7. Explain why good stewardship of Earth is a biblical concept.
8. Explain why it is important to use renewable natural resources wisely, even though they can be replenished over time.
9. Write a paragraph about a way in which some part of the creation has been degraded in or near your community. Is this problem becoming better, worse, or staying about the same?

Chapter 4

Matter and Minerals

The universe God created is a place of astonishing beauty—beauty you can see clearly in this crystal of rhodochrosite, in a flower of the field, or the stars on a dark night. Rhodochrosite crystals such as this are prized by mineral collectors, but the stone is not used as a gemstone because of its lack of hardness and the ease with which it breaks (or cleaves) along certain planes. Hardness and cleavage are two mineral properties you will learn about in this chapter.

The minerals of Earth are not only a delight to our eyes, but a resource used in many of the materials we need in today's complex society. People have dug in Earth for thousands of years, searching for precious metals, beautiful gems, and industrial materials, such as copper and iron. Will we run out of the minerals we need? How do we balance our need to mine Earth's mineral riches with the need to protect land and water resources? These are among the questions Earth scientists seek to address.

Objectives

After studying this chapter and completing the exercises, you should be able to do each of the following tasks, using supporting terms and principles as necessary.

1. Describe the structure of atoms.
2. Distinguish between elements, atoms, compounds, and molecules.
3. Describe ionic bonds, covalent bonds, and metallic bonding.
4. Use the five-part definition of *mineral* to classify Earth materials as mineral or non-mineral.
5. Describe ways in which minerals form.
6. Use a mineral's chemical formula to identify the group a mineral belongs to.
7. Identify mineral samples based on their physical properties.
8. Explain the difference between a mineral and an ore.

Vocabulary Terms

You should be able to define or describe each of these terms in a complete sentence or paragraph.

1. alteration
2. atom
3. atomic number
4. carbonate
5. cleavage
6. compound
7. covalent bond
8. crystal
9. crystal form
10. crystallization
11. density
12. electron
13. element
14. fracture
15. gem
16. halide
17. hardness
18. ionic bond
19. lava
20. luster
21. magma
22. metallic bonding
23. mineral
24. Mohs hardness scale
25. native element
26. neutron
27. ore
28. oxide
29. placer deposit
30. proton
31. rock
32. silicate
33. streak
34. sulfate
35. sulfide

4.1 Atoms, Elements, and Crystals

Every substance in Earth is composed of the elements listed in the Periodic Table of the Elements. An *element* is a substance that cannot be broken down into simpler substances by physical or chemical means. As an example, lead sulfide (PbS), also known as galena, is not an element because it can be broken down into the elements lead (Pb) and sulfur (S). There are 90 naturally occurring elements in the Earth. Scientists have created an additional 28 elements in sophisticated laboratories, bringing the total number of known elements to 118.

4.1.1 Atomic Structure

The smallest individual particle of an element is an *atom*. Atoms are composed of three smaller subatomic particles: *protons*, *neutrons*, and *electrons*. At the center of each atom is a nucleus containing the atom's protons and neutrons. These two subatomic particles possess roughly the same mass, but protons have a positive (+) electrical charge, whereas neutrons have no charge.

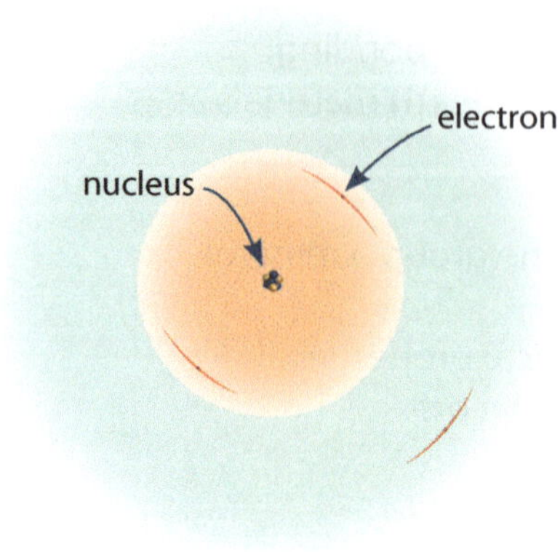

Figure 4.1. A representation of an atom of lithium. The nucleus contains three protons (blue) and four neutrons (yellow). Outside the nucleus are the atom's three electrons—shown as streaks to suggest their motion as well as their strange character.

Elements are distinguished by the number of protons their atoms contain in the nucleus. Every hydrogen atom has one proton in its nucleus, every lithium atom has three protons in its nucleus, and every iron atom has 26 protons in its nucleus. The number of protons an element has in the nuclei[1] of its atoms is called the *atomic number* of that element. Figure 4.1 is a representation of an atom of the element lithium. The nucleus of this atom contains three protons and four neutrons, and there are three electrons outside the nucleus.

For a given element, the number of protons in the atomic nucleus is fixed. But the atoms of most elements have some variability in the number of neutrons they have in their nuclei. In nature, at-

Atomic Number	Element	Symbol	Number of Protons	Number of Neutrons	Number of Electrons
1	hydrogen	H	1	0 or 1	1
2	helium	He	2	1 or 2	2
3	lithium	Li	3	3 or 4	3
4	beryllium	Be	4	5	4
5	boron	B	5	5 or 6	5
6	carbon	C	6	6 or 7	6
7	nitrogen	N	7	7 or 8	7
8	oxygen	O	8	8, 9, or 10	8
9	fluorine	F	9	10	9
10	neon	Ne	10	10, 11, or 12	10
11	sodium	Na	11	12	11
12	magnesium	Mg	12	12, 13, or 14	12

Table 4.1. Atomic numbers and subatomic particles for the first 12 elements.

1 The plural of *nucleus* is *nuclei*.

oms of oxygen always have eight protons in their nuclei—that is what makes these atoms oxygen—but an oxygen atom can have 8, 9, or 10 neutrons.

Electrons are tiny particles possessing negative (–) electrical charge and contained in different regions around the nucleus. (The electrons in Figure 4.1 are shown as streaks because they are moving very fast, and because they aren't simply "particles"—they exhibit spread-out, wave-like characteristics. There are many things about electrons we do not understand.) In an ordinary atom, the number of electrons equals the number of protons. In an atom of carbon, the six protons, each with a charge of +1 for a total charge of +6, are balanced out by the six electrons, each with a charge of –1 for a total charge of –6. Electrons are much less massive than protons or neutrons, having about 1/2000th of the mass of a proton.

Table 4.1 lists the twelve lightest elements—those with the lowest atomic numbers—along with data about their subatomic particles. Note that the number of electrons in an atom of an element equals the number of protons, which equals the atomic number.

4.1.2 The Periodic Table of the Elements

The Periodic Table of the Elements is shown in Figures 4.2 and 4.3. The elements appear in the periodic table sequentially according to their atomic numbers, beginning with hydrogen in the upper left corner.

As illustrated in Figure 4.2, the metals occupy the left side and middle of the periodic table, while the nonmetals occupy the far right side. In between these is a small group of elements known as the metalloids. Most metals exhibit properties such as shiny metallic appearance, high electrical conductivity, and malleability. When a metal is malleable, it changes its shape when struck with a hammer rather than breaking or shattering. Nonmetal solids have a dull appearance, do not conduct electricity well, and are brittle—they break or crumble when hit with a ham-

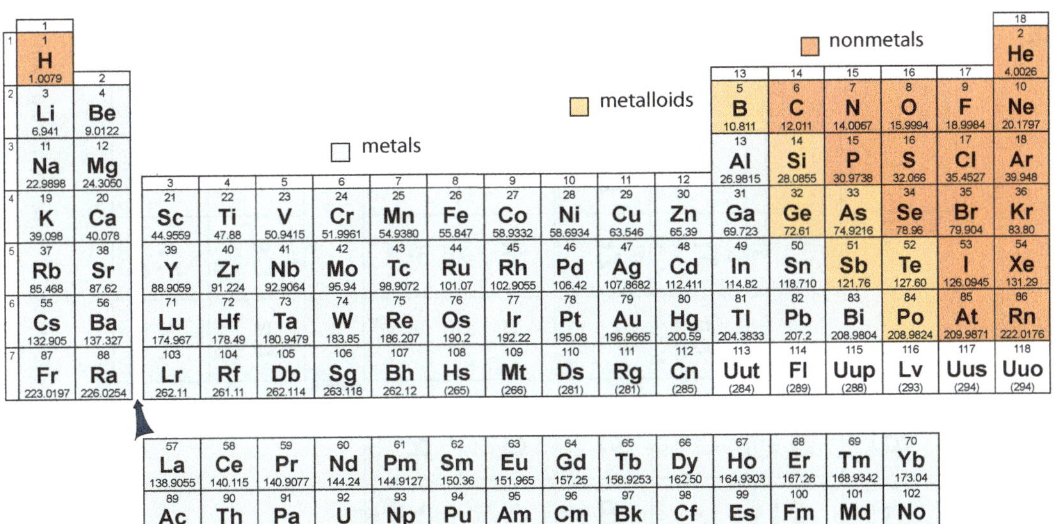

Figure 4.2. Metals, nonmetals, and metalloids in the periodic table.

Periodic Table of the Elements

■ most common elements in Earth's crust ■ not found in nature ■ other elements

Key: group (column) 1; period 1; atomic number 1; chemical symbol H; element name Hydrogen; atomic mass 1.0079

Period	1	2	3	4	5	6	7	8	9	10	11	12	13	14	15	16	17	18
1	1 H Hydrogen 1.0079																	2 He Helium 4.0026
2	3 Li Lithium 6.941	4 Be Beryllium 9.0122											5 B Boron 10.811	6 C Carbon 12.011	7 N Nitrogen 14.0067	8 O Oxygen 15.9994	9 F Fluorine 18.9984	10 Ne Neon 20.1797
3	11 Na Sodium 22.9898	12 Mg Magnesium 24.3050											13 Al Aluminum 26.9815	14 Si Silicon 28.0855	15 P Phosphorus 30.9738	16 S Sulfur 32.066	17 Cl Chlorine 35.4527	18 Ar Argon 39.948
4	19 K Potassium 39.098	20 Ca Calcium 40.078	21 Sc Scandium 44.9559	22 Ti Titanium 47.88	23 V Vanadium 50.9415	24 Cr Chromium 51.9961	25 Mn Manganese 54.9380	26 Fe Iron 55.847	27 Co Cobalt 58.9332	28 Ni Nickel 58.6934	29 Cu Copper 63.546	30 Zn Zinc 65.39	31 Ga Gallium 69.723	32 Ge Germanium 72.61	33 As Arsenic 74.9216	34 Se Selenium 78.96	35 Br Bromine 79.904	36 Kr Krypton 83.80
5	37 Rb Rubidium 85.468	38 Sr Strontium 87.62	39 Y Yttrium 88.9059	40 Zr Zirconium 91.224	41 Nb Niobium 92.9064	42 Mo Molybdenum 95.94	43 Tc Technetium 98.9072	44 Ru Ruthenium 101.07	45 Rh Rhodium 102.9055	46 Pd Palladium 106.42	47 Ag Silver 107.8682	48 Cd Cadmium 112.411	49 In Indium 114.82	50 Sn Tin 118.710	51 Sb Antimony 121.76	52 Te Tellurium 127.60	53 I Iodine 126.9045	54 Xe Xenon 131.29
6	55 Cs Cesium 132.905	56 Ba Barium 137.327	71 Lu Lutetium 174.967	72 Hf Hafnium 178.49	73 Ta Tantalum 180.9479	74 W Tungsten 183.85	75 Re Rhenium 186.207	76 Os Osmium 190.2	77 Ir Iridium 192.22	78 Pt Platinum 195.08	79 Au Gold 196.9665	80 Hg Mercury 200.59	81 Tl Thallium 204.3833	82 Pb Lead 207.2	83 Bi Bismuth 208.9804	84 Po Polonium 208.9824	85 At Astatine 209.9871	86 Rn Radon 222.0176
7	87 Fr Francium 223.0197	88 Ra Radium 226.0254	103 Lr Lawrencium 262.11	104 Rf Rutherfordium 261.11	105 Db Dubnium 262.114	106 Sg Seaborgium 263.118	107 Bh Bohrium 262.12	108 Hs Hassium (265)	109 Mt Meitnerium (266)	110 Ds Darmstadtium (281)	111 Rg Roentgenium (281)	112 Cn Copernicium (285)	113 Uut Ununtrium (284)	114 Fl Flerovium (289)	115 Uup Ununpentium (288)	116 Lv Livermorium (293)	117 Uus Ununseptium (294)	118 Uuo Ununoctium (294)

57 La Lanthanum 138.9055	58 Ce Cerium 140.115	59 Pr Praseodymium 140.9077	60 Nd Neodymium 144.24	61 Pm Promethium 144.9127	62 Sm Samarium 150.36	63 Eu Europium 151.965	64 Gd Gadolinium 157.25	65 Tb Terbium 158.9253	66 Dy Dysprosium 162.50	67 Ho Holmium 164.9303	68 Er Erbium 167.26	69 Tm Thulium 168.9342	70 Yb Ytterbium 173.04
89 Ac Actinium 227.0278	90 Th Thorium 232.0381	91 Pa Protactinium 231.0359	92 U Uranium 238.0289	93 Np Neptunium 237.0482	94 Pu Plutonium 244.0642	95 Am Americium 243.0614	96 Cm Curium 247.0703	97 Bk Berkelium 247.0703	98 Cf Californium 251.0796	99 Es Einsteinium 252.083	100 Fm Fermium 257.0951	101 Md Mendelevium 258.10	102 No Nobelium 259.1009

Figure 4.3. The Periodic Table of the Elements, highlighting in red the eight elements that make up almost 99% of Earth's crust. Elements that do not exist in nature on Earth are shown in gray.

mer. The metalloids have properties that are not clearly metallic or nonmetallic, but can appear somewhat metallic or somewhat nonmetallic depending on conditions.

Elements with similar chemical properties are grouped together in the periodic table in columns known as *groups*. The metals in Group 1, such as lithium, sodium, and potassium (Li, Na, and K), are all highly reactive, lightweight metals. Because of this, you will never find a piece of one of these metals in nature. There are nuggets of gold in nature because gold is fairly unreactive, but there are no nuggets of sodium or potassium. These metals react instantly with any oxygen or water they come in contact with. At the other end of the periodic table, the elements in Group 18, such as helium, neon, and argon (He, Ne, and Ar) are all extremely unreactive (inert) gases.

Most of Earth's crust is composed of just eight elements, which are highlighted in red in Figure 4.3. The abundance of these elements in Earth's crust is illustrated in the pie chart in Figure 4.4. You might be surprised to see that the most abundant element in the crust is oxygen, an element we usually think of as a gas. Most of this crustal oxygen is chemically combined with silicon to form quartz (SiO_2) and other compounds. Silicon is the second most abundant element, followed by aluminum, iron, calcium, magnesium, sodium, and potassium. Eighty-two elements fit in the "all others" category, and most of these are scarce. Gold (Au) is one of the scarcest elements; only about one out of every one billion (1,000,000,000) atoms in the Earth's crust is an atom of gold.

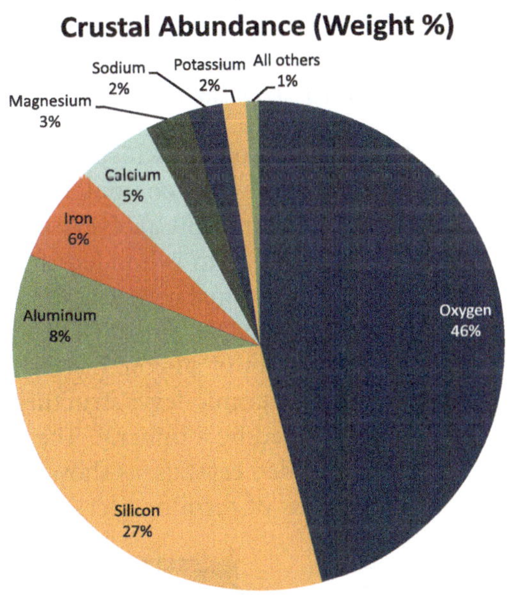

Figure 4.4. About 99% of Earth's crust is composed of just eight elements. Oxygen (46%) is by far the most abundant of these elements, followed by silicon (27%).

4.1.3 Compounds and Crystals

Almost all atoms in Earth's crust are chemically bonded to other atoms; relatively few exist as free atoms. A *compound* is a substance made of two or more elements that have chemically bonded to each other. Water is an example of a compound in which the atoms form molecules, identical clusters of atoms chemically bonded together. In water (H_2O), each molecule is composed of two hydrogen atoms bonded to a single oxygen atom. Some other molecular compounds are carbon dioxide (CO_2), methane (CH_4), and ammonia (NH_3).

Pyrite is another example of a compound. Its formula, FeS_2, indicates that in pyrite there is one iron atom for every two atoms of sulfur. You will not find individual molecules of FeS_2. Instead, pyrite exists as *crystals* composed of large numbers of atoms.

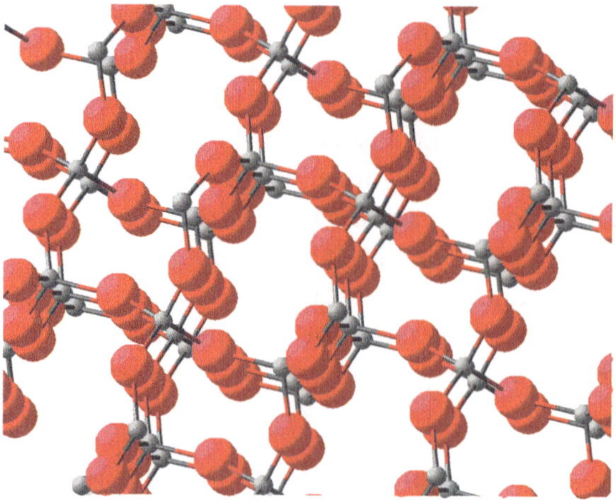

Figure 4.5. The atomic structure of a crystal of quartz. Oxygen atoms are red; silicon atoms are gray.

Figure 4.6 The hexagonal (six-sided) structure of quartz crystals is due to their internal atomic structure.

A crystal is a solid in which atoms are arranged in orderly, repeating patterns, such as exhibited by the computer model of a quartz crystal shown in Figure 4.5. In a crystal of quartz (SiO_2), every silicon atom is bonded to four oxygen atoms, and every oxygen atom is bonded to two silicon atoms. A single crystal of quartz weighing just six grams contains about 120,000,000,000,000,000,000,000 atoms of oxygen and 60,000,000,000,000,000,000,000 atoms of silicon—two atoms of oxygen for every atom of silicon. (These numbers indicate just how incredibly small atoms are!) The atomic-level structure shown in Figure 4.5 determines the shape of crystals formed by a mineral. The quartz crystal structure shown in Figure 4.5 results in six-sided crystals, as shown in Figure 4.6.

The process of forming a crystal is called *crystallization*. A familiar example of crystallization is the freezing of water, which happens when water cools down to the point where the individual water molecules are moving slowly enough that electrical attractions between molecules cause the molecules to attach to each other. Water crystallizes at 0°C. By contrast, the mineral quartz crystallizes at about 600°C.

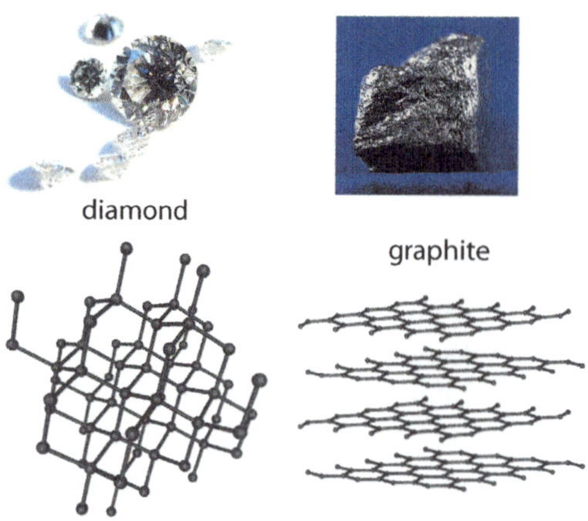

Figure 4.7. Both diamond and graphite are made out of pure carbon. In diamond, carbon atoms are arranged in a three-dimensional network. In graphite, the carbon atoms are arranged in weakly bonded sheets that can easily slide past each other.

Sometimes, atoms can be arranged in more than one way to make different crystals. Carbon has two primary forms in nature: diamond and graphite. Diamond can be fashioned into beautiful gemstones, and is known as one of the hardest—that is, most scratch-resistant—substances in existence. Graphite is a substance that seems to be the opposite of diamond in many ways. Graphite is a dark gray, slippery-feeling substance that is one of the softest (i.e., least scratch resistant) substances known. Graphite is the primary ingredient in pencil lead.[2] Both diamond and graphite are made of pure carbon (C), but the atoms in crystals of these two substances are arranged differently, as Figure 4.7 shows. In a crystal of diamond, carbon atoms are arranged in a three-dimensional network, with strong chemical bonds that hold the atoms together in all directions. In graphite, atoms are combined in two-dimensional sheets. There are strong chemical bonds within individual sheets, but very weak bonds between graphite sheets. This weakness between individual sheets of graphite is actually what makes graphite so useful as pencil lead. When you rub pencil lead across a sheet of paper, the graphite layers peel off the crystal and cling to the sheet of paper.

In some cases, compounds form crystals that have visible geometric forms, such as the crystal of calcite shown in Figure 4.8. However, most of the time in nature crystals do not show these kinds of distinct crystal faces; sometimes the crystal shapes of specimens are only apparent with a microscope. But even if no crystal faces are apparent, if the atoms in the substance are arranged in an orderly, geometrical structure the specimens are crystals.

Figure 4.8. Sometimes the mineral calcite ($CaCO_3$) makes beautiful crystals, but more often its crystalline structure can only be seen at the microscopic level.

4.1.4 Chemical Bonds

A chemical bond is a strong attachment between the atoms in a molecule or crystal.[3] Chemical bonds hold the hydrogen and oxygen atoms together in a water molecule, and the sodium and chlorine atoms together in a crystal of sodium chloride (table salt). To help establish the difference between the basic structures exhibited by substances in nature, we will look briefly at the three types of chemical bonds.

In the *ionic bond*, electrons are *transferred* from metal atoms to nonmetal atoms, forming crystals in the process. When atoms gain or lose electrons, they become *ions*, atoms that have a net (i.e., unbalanced) electrical charge. When metal atoms ionize, they lose electrons to become positive (+) ions. When nonmetal atoms ionize, they gain electrons to become negative (–) ions. For example, sodium ionizes by losing one electron. The symbol for the sodium ion is the chemical symbol for sodium, Na, with a plus sign added to signify its net charge: Na^+. Similarly, the

2 Pencil lead does not contain any of the element lead, Pb.

3 More generally, there are also weak chemical bonds between molecules that cause molecules to cling together in molecular solids and liquids. As important as these kinds of bonds are, they are not our focus here.

symbol for a chloride[4] ion is the chemical symbol for chlorine, Cl, with a negative sign added: Cl^-. The ionic bond that holds ions together is an electrical attraction, similar to the static electrical attraction that can hold a balloon to a wall, or socks to other pieces of clothing when removed from a hot clothes dryer.

A *covalent bond* occurs when two nonmetal atoms *share* pairs of electrons (rather than transferring electrons from one atom to another). The water (H_2O), methane (CH_4), carbon dioxide (CO_2), and ammonia (NH_3) molecules are each held together by covalent bonds between atoms in the molecules.

It is important to note that the distinction between ionic and covalent bonding, like all scientific reasoning, is a scientific way of *modeling* the chemical behavior we see in nature. However, this either-or way of looking at chemical bonding is oversimplified. The bonds in most substances exhibit both ionic and covalent character. Crystals like sodium chloride are mostly ionic in character; molecules like methane are mostly covalent in character. But in some substances, both ionic and covalent characters are present. The different forms of silicon dioxide, SiO_2, are like this. Quartz and glass are both forms of silicon dioxide. Quartz (Figures 4.5 and 4.6) is a crystal because of the way the atoms are arranged in a regular geometric pattern. But the chemical bonds between silicon and oxygen atoms possess less than 50% ionic character and are thought of as mostly covalent. For this reason, silicon and oxygen can also form networks of bonds that are *amorphous*, which means they are arranged in a large network that does not have a regular geometric pattern, illustrated in Figure 4.9. This is the way the atoms in glass are structured. Since there is no regular crystalline pattern, the best way to think of amorphous SiO_2 is as one huge molecule. However,

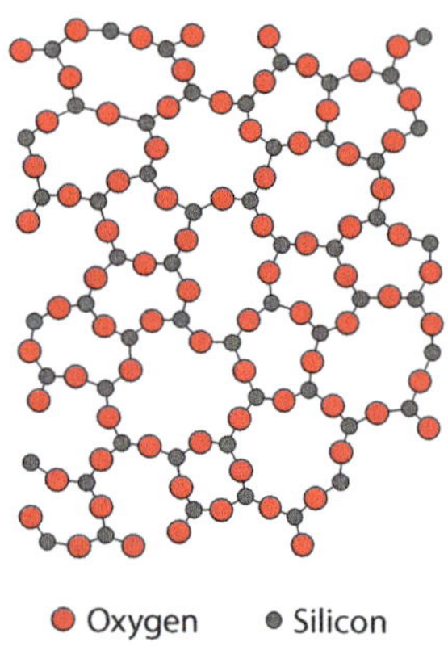

Figure 4.9. The amorphous arrangement of oxygen and silicon atoms in glass.

Type of Bond	Types of Atoms Bonded	Type of Compound Formed	Examples
Ionic	metals to nonmetals	Crystal	NaCl—halite (table salt) FeS_2—pyrite CaF_2—fluorite
Covalent	nonmetals to nonmetals	Molecule	H_2O—water CH_4—methane SiO_2—quartz
Metallic	metal atoms within a pure metal	Metal	Au—gold Cu—copper

Table 4.2. Types of chemical bonds.

4 Negative ions are named by changing the ending of the element to "*-ide*." Thus, oxygen atoms become oxide ions, chlorine atoms become chloride ions, and fluorine atoms become fluoride ions.

in all SiO_2 structures, the bonds are primarily covalent in character, so we think of the electrons involved in the bonding as shared between the oxygen and silicon atoms, rather than as belonging specifically to either.

The third type of chemical bond is *metallic bonding*,[5] which forms between the atoms in a pure metal. In metallic bonding, the outermost electrons of the atoms can be pictured as freely floating between the atoms of the metal. These free electrons—called conduction electrons—are responsible for the property of high electrical conductivity exhibited by metals.

Table 4.2 summarizes these three types of chemical bonds.

Learning Check 4.1

1. Describe an atom of helium, including the location and number of its subatomic particles.
2. Categorize the following elements as metal or nonmetal: carbon, oxygen, sodium, magnesium, aluminum, silicon, chlorine, iron.
3. How many elements are among those composing 99% of Earth's crust? Which of these are the first and second most abundant?
4. Describe ionic bonds, covalent bonds, and metallic bonding, and list the types of elements involved in each.

4.2 Minerals

The crust of Earth contains thousands of different chemical compounds in various states. There are liquids, such as water and petroleum. There are gases, such as the oxygen and nitrogen in spaces in soil, or the carbon dioxide and sulfur dioxide seeping out of Earth in volcanic regions, illustrated in Figure 4.10. And Earth's crust contains various solids, such as rocks and, in some places in polar or alpine regions, ice.

Figure 4.10. Volcanic gases escaping from near the summit of a volcano in Washington. The fumes are not a mineral, because they are gaseous, but the yellow sulfur deposited by the gases is a mineral.

4.2.1 Definition of Mineral

With most rocks, if you look closely you will see that they are composed of individual grains of matter. The rock shown in Figure 4.11 is composed primarily of just three minerals: quartz, orthoclase feldspar, and biotite mica. A *mineral* is a

5 The nature of metallic bonding is such that large numbers of metal atoms are always involved. For this reason, I use the term *bonding* instead of simply *bond*. Ionic and covalent bonds exist between individual atoms, but metallic bonding always involves a host of atoms.

naturally occurring, inorganic, crystalline solid with a definite chemical composition and definite physical properties (discussed in Section 4.3). There is a lot in that definition, so let's take a closer look at each part.

Figure 4.11. Rocks are composed of minerals. This polished piece of granite contains orthoclase feldspar (pink), quartz (gray), biotite mica (black), and minor amounts of other minerals.

- *Naturally occurring*—In order to be classified as a mineral, a substance must have formed by natural processes. A diamond mined from Earth's crust is a mineral because it formed naturally. A synthetic diamond, created by squeezing pure carbon under millions of pounds per square inch of pressure, is not a mineral because it did not form naturally.
- *Inorganic*—A mineral must be *inorganic*. This means that minerals form from non-living material rather than from the decay of once-living organisms. (*Organic* materials are carbon-containing substances that are or were part of living organisms.) Many minerals form by crystallization from molten rocks, such as lava. This process is clearly inorganic; it does not involve biological processes. By this definition, the materials that make coal are not minerals because they form by the transformation of plant material.
- *Crystalline solid*—Minerals are crystalline solids. For example, the sodium ions (Na^+) and chloride ions (Cl^-) in a sodium chloride crystal are arranged in a three-dimensional grid with the ions lined up in rows that are perpendicular (90° angle) to each other. This regular arrangement of ions is reflected in the shape of crystals of many minerals. The sodium chloride mineral is called halite, shown in Figure 4.12.
- *Definite chemical composition*—Minerals have definite chemical compositions, reflected in chemical formulas such as SiO_2 (quartz), $CaCO_3$ (calcite), and CaF_2 (fluorite). Sometimes elements can combine in more than one ratio. This enables different minerals composed of the same elements to exist. For example, iron and oxygen can combine to form hematite (Fe_2O_3) or magnetite (Fe_3O_4). Because these two substances have

Figure 4.12. These cubic crystals of halite (NaCl) reflect the orderly, perpendicular arrangement of sodium and chloride ions in its internal structure.

different chemical compositions, they are different minerals, with different properties.

A *rock* is a solid mixture of one or more minerals. Some rocks consist entirely of one mineral. Limestone, for instance, is composed of one mineral, calcite, though it often contains minor amounts of other minerals as impurities. The piece of granite in Figure 4.11 is composed mostly of quartz, feldspar, and biotite mica, but it also contains a number of other minerals that make up less than one percent of the rock.

Figure 4.13. Seawater can be evaporated in small ponds. The salt that remains is then collected and sold.

4.2.2 Formation of Minerals

Minerals form in Earth's crust by a variety of means. One of the easiest mineral-forming processes to visualize is *evaporation*. Seawater, as mentioned in Chapter 3, is about 3.5% salt, mostly sodium chloride (NaCl). If a quantity of seawater is isolated from the rest of the ocean, as shown in the evaporation ponds in Figure 4.13, the water will evaporate and leave salt behind. This happens naturally in deserts, such as the salt flat shown in Figure 4.14. In addition to halite, a variety of other minerals can form by evaporation from seawater or mineral-rich water from lakes in arid regions.

Figure 4.14. When water evaporates from desert lakes, it can leave layers of salt behind.

Most of the rocks in Earth's crust were formed by *crystallization* from *magma* or *lava*. Magma is the term for molten rock inside of Earth; the term lava applies to molten rock when it comes out of Earth, such as at the erupting volcano shown in Figure 4.15. As magma or lava cools, various minerals begin to crystallize. Minerals formed as magma cools have much higher freezing points than water does: typically in the range of 600 to

Figure 4.15. When lava cools and solidifies, it forms various minerals by crystallization.

1200°C. When water freezes, it forms just one substance, ice. Because magma contains a number of different elements, it "freezes" into a mixture of different minerals, as we saw in Figure 4.11.

A third way minerals form is through *alteration*, a process by which a mineral is transformed from one mineral to another by the addition or subtraction of elements, typically transported by fluids. Alteration can occur at or near Earth's surface, where a variety of processes break down minerals in rock to create soil. Feldspar and mica grains near the surface can be transformed into various types of clay minerals, illustrated in Figure 4.16. Deep in Earth's crust, heat, pressure, and hot fluids move elements in and out of mineral grains, transforming the clay minerals back into mica. These are just two of the many examples of minerals that can be formed by alteration.

Figure 4.16. When granite weathers, some of the minerals are altered to clay as the rock crumbles. The pocket knife is in the photo to give an idea of the size of the mineral grains.

The mineral-forming processes we have looked at, as well as others, are often studied through experiments. An obvious example is the formation of halite (NaCl) through evaporation, a process that is easily modeled with table salt and equipment found in any kitchen. The crystallization of minerals from magma is more difficult to model, but is an experiment that has been carried out by geologists a number of times. Geologists have melted rocks and slowly cooled the liquid while observing which minerals crystallize first, second, third, and so forth, until the entire sample has solidified.

4.2.3 Mineral Groups

There are over 4,500 different minerals found in Earth's crust, but most of the crust is composed of just a few dozen of them. These minerals can be classified into various mineral groups, which are categories of minerals based on chemical composition.

The most abundant mineral group is the *silicates*. The silicates are minerals that contain silicon (Si) and oxygen (O). Most of Earth's crust and mantle is composed of silicate minerals, a result of the fact that oxygen and silicon are the most common elements in the crust and mantle—Figure 4.4. The most basic silicate mineral is quartz, which has the formula SiO_2, as we have seen. Other silicate minerals incorporate metal ions—such as aluminum, potassium, calcium, and iron—into their crystal structures. Feldspar minerals, such as orthoclase and plagioclase, contain potassium, sodium, calcium, and aluminum ions, and make up about half of Earth's crust. In silicate minerals, the silicon and oxygen atoms are bonded to each other with covalent bonds, and the metal ions are bonded to the silicon-oxygen framework by ionic bonds.

Other common silicate minerals include the mica minerals biotite and muscovite (which separate easily into sheets), and pyroxene, amphibole (pronounced AM-fi-bowl), and olivine (most abundant in oceanic rocks formed from magma and lava). Several of these silicate minerals are shown in Figure 4.17.

Most other mineral groups are scarce compared to the silicates, but many minerals within these groups are economically important as material resources.

- *Native elements* are minerals composed of just one element. Examples include gold (Au), copper (Cu), graphite (C), silver (Ag), and sulfur (S). Only a handful of elements occur as native elements.
- *Carbonates* are minerals that contain carbon and oxygen covalently bonded to each other to form carbonate ions. These minerals can be identified by the CO_3 in their chemical formulas. Limestone is composed mostly of the mineral calcite,

Figure 4.17. Some common silicate minerals.

$CaCO_3$. Limestone is a widespread rock in the upper part of Earth's crust. Carbonates are the second-most abundant group of minerals in Earth's crust, after the silicates.

- *Sulfides* are minerals that contain a metal ion bonded to sulfur, but do not contain oxygen. Many important ore minerals are sulfides, including pyrite, commonly called "fool's gold" because of its goldish color, and galena, an important source of lead. Sulfide-group minerals have a metallic appearance.
- *Sulfates* are minerals containing the sulfate ion, which appears in the chemical formula as SO_4. A widespread example is gypsum.
- *Oxides* are minerals that contain a metal ion bonded to oxygen. Some oxides are important sources for metals used in industry, such as iron from hematite, and chromium from chromite.
- *Halides* are minerals in which a metal ion bonds to an element from Group 17 of the periodic table—fluorine, chlorine, bromine, or iodine. The most familiar halide mineral is halite (Figure 4.12). Another important halide is fluorite, which is mined for a variety of industrial purposes.

The silicate and non-silicate mineral groups are summarized in Table 4.3, and several non-silicate minerals are shown in Figure 4.18.

Group	Chemical Formula	Examples
Silicates	includes Si and O	quartz—SiO_2 orthoclase—$KAlSi_3O_8$ olivine—$(Mg,Fe)_2SiO_4$
Native elements	a single element	gold—Au copper—Cu diamond—C
Carbonates	includes carbonate (CO_3)	calcite—$CaCO_3$ rhodochrosite—$MnCO_3$
Sulfides	includes S but not O	pyrite—FeS_2 galena—PbS bornite—Cu_5FeS_4
Sulfates	includes sulfate (SO_4)	gypsum—$CaSO_4 \cdot 2H_2O$
Oxides	includes O but not CO_3 or SO_4	hematite—Fe_2O_3 magnetite—Fe_3O_4 corundum—Al_2O_3
Halides	includes F, Cl, Br, or I	halite—NaCl fluorite—CaF_2

Table 4.3. Mineral Groups.

Figure 4.18. Some important non-silicate minerals.

4.2.4 Gems

Gems (or gemstones) are scarce minerals that are highly valued for their exceptional beauty. Most gems are dull and relatively unattractive in their natural state, and must be cut and polished to reveal their beauty. The sapphire in Figure 4.19 has not been cut and does not have the flat surfaces (facets) you might see on a gemstone in a jewelry store. The more scarce and valuable gems, such as diamonds and emeralds, are known as *precious* gems, and the less valuable gems are known as *semi-precious* gems.

Figure 4.19. An uncut sapphire.

Figure 4.20. Three precious gemstones: diamond (left), ruby (center), and sapphire (right).

Figure 4.21. Three quartz gemstones: amethyst (left), citrine (center), and smoky quartz (right).

	Gem	Mineral Name	Valuable Colors
Precious	diamond	diamond	colorless, yellow
	emerald	beryl	green
	opal	opal	brilliant hues
	ruby	corundum	red
	sapphire	corundum	blue, pinkish-orange
Semi-precious	amethyst	quartz	purple
	citrine	quartz	yellow
	smoky quartz	quartz	brown
	garnet	garnet	red, green, orange
	jade	jadeite or nephrite	green
	peridot	olivine	green
	tourmaline	tourmaline	red, blue-green, green, pink
	turquoise	turquoise	greenish-blue

Table 4.4. Important gems.

Most gems are actually common minerals that have some chemical impurity that gives them color and makes them valuable. For example, the mineral corundum is a hard, relatively unattractive mineral commonly used as an abrasive, such as on sandpaper. However, if corundum contains tiny amounts of chromium, it has a vivid red color, and is called ruby. If corundum contains trace amounts of titanium, iron, or copper, it displays a wide variety of other colors, and is known as sapphire. Figure 4.20 shows some beautifully cut precious gems.

The small size or imperfections of many diamonds makes them unsuitable for use as gemstones. Like corundum, non-gem diamonds are used for industrial purposes. Many dental drills have tiny diamonds on them for grinding or polishing teeth.

Quartz, as we have seen, is one of the most common minerals in Earth's crust. Quartz is ordinarily colorless or white, but impurities or imperfections in the crystal structure gives colors to quartz that make it a semi-precious gemstone. Purple quartz is known as amethyst, yellow quartz is called citrine, and brownish-gray to black quartz is called smoky quartz. These three semi-precious gems are shown in Figure 4.21.

The size of precious gems is measured in *carats*, a non-metric system unit of mass. A carat is equal to 200 milligrams, and a one-carat diamond has a diameter of about 6 millimeters.[6] Table 4.4 lists some important gems.

Learning Check 4.2

1. Classify each of the following substances as mineral or non-mineral: quartz, steam, aluminum can, groundwater, feldspar, tree roots in soil, clay in soil, decomposed leaves in soil.
2. What is the difference between a mineral and a rock?
3. Identify mineral group to which each of the following minerals belongs: graphite (C), stibnite (Sb_2S_3), pyrolusite (MnO_2), argentite (AgS_2), cassiterite (SnO_2), sylvite (KCl), strontianite ($SrCO_3$), wollastonite ($CaSiO_3$), anhydrite ($CaSO_4$), sillimanite (Al_2SiO_5).
4. What is the difference between a) ordinary corundum and b) rubies and sapphires?
5. Describe three semi-precious varieties of quartz.

4.3 Mineral Properties and Identification

Geologists and mineral collectors often have to identify unknown mineral samples. There are thousands of minerals, but with experience, and knowledge of

6 Don't confuse *carat* and *karat*, which are pronounced the same. The carat is a measurement of gem size. The karat is a measurement of the purity of gold. Pure gold is 24 karat.

the properties of various minerals, it is usually possible to identify the more widespread minerals without using sophisticated laboratory techniques.

We cannot see the elements in a mineral just by looking at it, nor can we see the mineral's internal atomic structure. Whether in the field or in a laboratory, geologists use basic properties that are easy to see or measure, such as color and hardness, to identify minerals.

4.3.1 Luster, Color, and Streak

The property of *luster* is due to the way a mineral reflects light off its surface. One of the most important distinctions is between minerals that have metallic luster—that is, those that reflect light like a piece of metal—and those that have non-metallic luster. Minerals with non-metallic luster are described with terms such as submetallic, glassy, pearly, silky, earthy, waxy, and dull. Three of these lusters are represented by the minerals shown in Figure 4.22.

Figure 4.22. Luster. Pyrite (left) has a metallic luster, topaz (center) has a glassy luster, and kaolinite (right) has a dull or earthy luster.

Another property is color—one of the first things we notice about a mineral. Some minerals have distinctive colors that play a key role in identification. For example, the mineral azurite, pictured in Figure 4.23, has a characteristic blue color. But color can be an unreliable property for identifying many minerals because impurities cause them to occur with a wide variety of colors. As already mentioned, quartz is normally white or colorless, but impurities give it a number of other colors.

Figure 4.23. Azurite has a distinctive blue color.

Streak is the color of a mineral when it is powdered. A mineral's streak tends to be much more consistent than its color. The color of the mineral hematite can vary from reddish-brown to black to silver, but its streak is always rust-colored. Streak is typically determined by rubbing a sample over an unglazed ceramic streak plate, as illustrated in Figure 4.24.

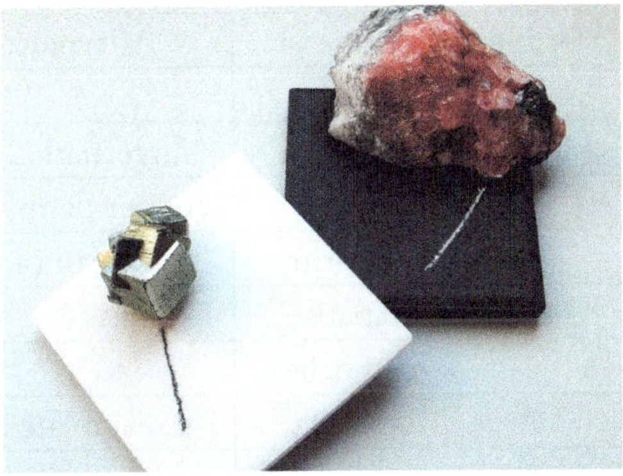

Figure 4.24. Ceramic streak plates, showing the dark brown streak of pyrite and the white streak of rhodochrosite.

4.3.2 Crystal Form, Hardness, Cleavage, and Density

Because atoms in a mineral are arranged in orderly patterns, they can form crystals that have distinctive shapes. *Crystal form* is the shape a crystal mineral will have if the crystal is free to form without being constrained. In most cases, when minerals form they are in places where they cannot grow freely, and obvious crystal faces do not form. But if a mineral forms in some sort of cavity, beautiful multi-faced crystals can form. The quartz crystals shown in Figure 4.25 grew in the cavity inside a rock, forming a wonderful object called a *geode*. The gypsum crystals in Figure 4.26 grew in a cave in Mexico that was discov-

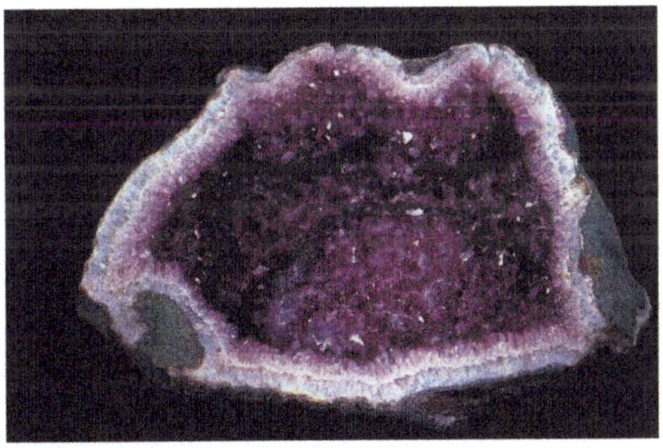

Figure 4.25. Amethyst crystals lining the inside of a geode.

Figure 4.26. Some of these gypsum crystals, which grew underwater in a cave in Mexico, are up to 12 meters long. Note the size of the person at lower right in the photo.

Hardness	Mineral	Alternates
1	talc	
2	gypsum	fingernail (2.5)
3	calcite	copper penny (3)
4	fluorite	iron nail (4.5)
5	apatite	glass (5.5)
6	orthoclase	
7	quartz	streak plate (7)
8	topaz	
9	corundum	
10	diamond	

Table 4.5 The Mohs hardness scale.

ered as miners were drilling to expand a silver-lead-zinc mine.

Hardness is a measure of a mineral's resistance to scratching. Hardness is one of the most useful properties for identifying mineral samples. The *Mohs scale* rates mineral hardness on a scale of 1 to 10, where a rating of 1 represents an extremely soft substance and 10 represents an extremely hard substance. The Mohs scale is shown in Table 4.5. The softest mineral on the Mohs scale is talc, which can easily be scratched with your fingernail, and is used, among other things, for making baby powder. The hardest mineral is diamond. Diamond will scratch any other mineral, and cannot itself be scratched by any substance (but it can be cut or shattered). For comparison, substances of comparable hardness are also listed in the table as alternates.

The hardness of a mineral is determined by comparing a sample to other minerals or objects with known hardness. If an unknown sample can scratch calcite (hardness = 3) but is itself scratched by apatite (hardness = 5), then the hardness of the unknown sample is about 4. Your fingernail has a hardness of 2.5, a penny has a hardness of 3, and glass has a hardness of 5.5. These three materials can be used to determine the hardness of many common minerals.

Most minerals will break if hit hard enough, even very hard minerals such as diamond or corundum. The major exception is metals such as gold and copper. As mentioned in Section 4.1.2, metals are malleable. Instead of shattering, metals will simply change shape when struck with a hammer. But in nonmetal crystals, chemical bonds within the crystal tend to be stronger in some directions and weaker in others. Because of this, some minerals are more likely to break in certain directions

Figure 4.27. Cleavage. Mica minerals (left) cleave into very thin sheets. Calcite (center) cleaves in three planes at angles of 55° and 74° to each other. Olivine (right) does not cleave, but fractures in curved lines as you can see in the lower center part of the image.

rather than others. *Cleavage* is the tendency of a mineral to break along certain planes. Mica, shown on the left in Figure 4.27, has a similar crystal structure to graphite (Figure 4.7). It cleaves along a single plane, making it possible to peel very thin sheets of mica from a large crystal. As you saw in Figure 4.12, halite breaks along three planes, perpendicular to each other. Calcite, shown in the center of Figure 4.27, also cleaves along three planes, but at angles of 55° and 74° relative to each other. Halite and calcite can both have white, semi-transparent crystals, so the difference in cleavage can be useful in distinguishing between the two.

Fracture occurs when minerals break without following the cleavage planes. Many minerals do not have planes of weakness in their crystal structure, and thus don't have a direction in which they are more likely to break. When a piece of quartz breaks, it shatters like a piece of glass rather than splitting along definite planes. Other minerals fracture irregularly, or in curved (conchoidal) lines, as exemplified by the piece of olivine shown on the right of Figure 4.27.

Minerals are also distinguished from one another by their *densities*. The density of a substance is its ratio of mass to volume, calculated using the formula

$$D = \frac{m}{V}$$

In this equation, D represents the density of a substance, m represents the mass of a sample, and V represents the volume of the same sample. If the mass is in units of grams and the volume is in units of cm^3, then the units of density will be g/cm^3. Minerals that contain elements with higher atomic numbers, such as uranium, lead, gold, and silver, tend to have high densities. Most of Earth's crust, however, is made of those eight elements shown in the pie chart in Figure 4.4. The relatively low atomic numbers of these elements means that their nuclei are small, which in turn means that the density of these elements is low. When minerals are composed of low-density elements, the density of the mineral tends to be low. However, the density of minerals also depends on how tightly the atoms are packed together.

4.3.3 Other Mineral Properties

Minerals have a number of additional physical and chemical properties that can help to identify unknown samples.

- *Magnetism*. A few minerals are slightly magnetic. The only common mineral that is strongly magnetic is magnetite, which can be used as a magnet. Figure 4.28 shows a sample of magnetite with steel nails stuck to it because of its magnetism.
- *Chemical reactivity*. Almost all minerals have specific chemical tests that can be used to help identify them. For example, a simple test for

Figure 4.28. Magnetite is magnetic, and small pieces of iron are attracted to it.

the mineral calcite is to put a drop of dilute hydrochloric acid on the mineral. Hydrochloric acid quickly reacts with calcite to produce carbon dioxide (CO_2) gas, which causes fizzing on the mineral surface.

- *Taste and smell.* Some minerals have distinctive tastes or smells. Halite (NaCl) tastes salty, which should come as no surprise. (However, licking mineral samples in a laboratory is definitely not a good idea!) Some sulfide minerals, such as pyrite (FeS_2), give off a rotten egg smell when heated or crushed. This is due to the formation of hydrogen sulfide gas, H_2S.

Learning Check 4.3

1. Which is a more reliable property to use for mineral identification, color or streak?
2. Why do some mineral samples show crystal form, while others do not?
3. Determine the approximate hardness of a mineral in each of these cases:
   a. A mineral sample can be scratched by apatite, but not by calcite.
   b. A mineral sample can be scratched by glass, but not by a nail.
   c. A mineral sample can be scratched by a nail, but not by a penny.
4. Describe what happens when a drop of hydrochloric acid is placed on calcite.

Answers to #3: a) 4 b) 4.5 c) 3.5

4.4 Mineral Resources

Except for energy from the sun, everything we use comes from Earth. Metals such as iron, copper, and chromium are extracted from ores that are mined from Earth. We use fuels from Earth, such as petroleum, natural gas, and coal, to provide energy for transportation and electrical power generation. Petroleum is also used to make a wide variety of plastics. Materials we use for construction, such as gravel, cement, and even timber, are all from Earth. The food we eat and the clothes we wear also come from Earth because all plants are dependent on soil and air. In this section we will look briefly at some of the factors related to our use of minerals as resources.

4.4.1 Ores

An *ore* is a mineral that contains a useful substance that can be mined at a profit. If a mineral deposit contains a low concentration of a metal such as copper, it will only be profitable to mine if the price of copper is high. If the price of copper is low, the deposit will not be classified as an ore.

Many elements, such as copper, platinum, silver, and gold, are scare in Earth's crust, but various geologic processes can make these elements more concentrated in certain places. For example, magma always contains at least some gold. Because gold atoms are quite heavy compared to the atoms that form most minerals, almost none of the gold atoms will become part of minerals as they crystallize from the magma. The last 1% of the magma to crystallize, therefore, will have a concentration of gold that is 100 times as great as the original magma. The concentration of gold in the last 0.01% of magma to crystallize will be 10,000 times greater than in the original magma. The last remnants of magma are often composed mostly of silica (SiO_2), water, and traces of other elements, such as gold. This hot fluid (referred to as a hydrothermal fluid) can squeeze into fractures in the surrounding rock to form quartz veins. Usually these veins do not contain gold, but occasionally they do.

4.29. Panning for gold from a placer deposit in a stream.

Another way gold is concentrated is by currents in streams. If nuggets or flakes of gold erode out of quartz veins, they can be washed down into streams. The density of gold is very high—nearly as high as the density of lead—so the water currents in streams tend to deposit the gold in certain places along gravel bars or bends in the stream. These stream mineral deposits are called *placer* (pronounced PLAH-sir) *deposits*. Figure 4.29 shows a gold pan, a tool used to separate dense mineral grains from less dense pieces of rock and mineral. Other dense minerals, such as diamonds and other gemstones, also concentrate in placer deposits. Much of the gold from the 1849 gold rush in California and the later gold rushes in Alaska and Yukon was extracted from placer deposits.

4.4.2 Mines

Minerals that are near Earth's surface can be removed from the ground by digging large, open pit mines, such as the one shown in Figure 4.30. Often, open pit mines are used to mine less-rich deposits of ore. Ores that are deeper in the crust can be obtained by underground mining. Some underground mines contain hundreds of miles of passages and extend more than a mile into Earth's crust.

4.30. An open pit copper mine in Utah. This pit is 4 kilometers (2.5 miles) across and almost a kilometer (0.6 miles) deep.

Mining, whether open pit or underground, often does a tremendous amount of damage to the environment. The land itself is scarred by the mining, and many of the waste materials produced as by-products of mining and ore processing poison water both on the surface and underground, as shown in Figure 4.31. The situation in this photo is an example of *acid mine drainage*, an environmentally devastating condition. Contaminated water in mining areas can contain dangerous levels of toxic substances such as lead, cadmium, and arsenic. More prudent methods that minimize such damage are now in use by responsible mining enterprises.

Figure 4.31. This stream has been contaminated by toxic metals from a mining area.

4.4.3 Minerals as Non-Renewable Resources

Although Earth's crust contains a tremendous amount of mineral resources, we usually think of these deposits as being somewhat limited and non-renewable. In other words, we could conceivably come to a point where some minerals are in short supply. Mines that are currently operating will sooner or later run out of ore, and new deposits will need to be found. It is likely that many of these new resources will be of lower quality or in more remote locations than mines that are active at present. Most mineral resources, however, will not become scarce in the next 100 years.

Perhaps the greatest challenge in the upcoming years will be mining in a way that minimizes negative impacts on human health and the environment. Mining provides jobs for people and raw materials for our society, but also has negative impacts on the well-being of humans, animals, and plants. It will require great wisdom for our society to make decisions on where and how to develop new mineral resources in ways that avoid contaminating the environment and are safe for the workers and local inhabitants.

Learning Check 4.4

1. Why is it that some deposits of the mineral bornite (Cu_5FeS_4) might be considered to be copper ores, but other deposits of the same mineral might not be considered to be copper ores?
2. Describe two ways that gold, which is scarce in Earth's crust, can be concentrated in nature to make ore deposits.
3. What are some environmental risks associated with mining?

Chapter 4 Exercises

Answer each of the questions below as completely as you can. Write your responses in complete sentences unless instructed otherwise.

1. Make a data table giving the atomic number, element symbol, element name, number of protons, and number of electrons for oxygen, silicon, potassium, iron, gold, and uranium.
2. Define the term *mineral.*
3. Describe three ways in which minerals can form.
4. Use the tables in Appendix A to identify the minerals described in the following table.

	Luster	Hardness	Color	Streak	Other
a	metallic	6	black	black	magnetic
b	nonmetallic	3	white	white	cleaves into cubes
c	nonmetallic	3	colorless	white	cleaves at angles
d	metallic	2.5	silver gray	black	cleaves into cubes
e	metallic	6.5	brassy yellow	green	uneven fracture
f	glassy	4	purple	colorless	cleaves in four directions
g	nonmetallic	6	pink	colorless	2 cleaves
h	nonmetallic	2.5	black	white	peels into very thin sheets

5. Calculate the density of a piece of native copper with a volume of 24 cm^3 and a mass of 214 g.
6. To identify a certain mineral sample, a geologist first weighs the sample and finds the mass to be 45.7 g. Next, the geologist measures the sample volume and finds it to be 12.5 cm^3. Determine the density of this mineral.
7. Comment on the following statement from the beginning of Section 4.4: "Except for energy from the sun, everything we use comes from Earth." Write a few paragraphs describing the sources of common and necessary things, such as food, clothing, building materials, and art supplies.
8. If Earth's crust contains a vast amount of the substances needed for our society to flourish, why do we still need to be careful stewards of mineral resources?

Answers
4. a) magnetite b) halite c) calcite d) galena e) pyrite f) fluorite g) orthoclase feldspar h) biotite mica
5. 8.9 g/cm^3
6. 3.66 g/cm^3

Experimental Investigation 2: Identifying Minerals

Overview

Although geologists use a variety of modern, and sometimes very expensive, tools to identify and analyze minerals, they also use the basic mineral properties you are learning in this course, especially when studying rocks in the field. In this lab, you will use luster, hardness, color, streak, and other mineral properties to identify mineral samples.

A kit containing various minerals and rocks.

Basic Materials List

- Mineral samples
- Streak plate
- Mineral hardness kit: copper penny, iron nail, small glass plate
- Magnet
- Dilute hydrochloric acid in small dropper bottle
- Magnifying glass
- Safety goggles and protective gloves

Safety and Other Guidelines

1. Because you will use hydrochloric acid to test minerals, it is necessary that you wear safety goggles and protective gloves during this laboratory exercise any time you are using the acid solution.
2. Only use the hydrochloric acid if you think the mineral might be calcite. One drop of hydrochloric acid solution is enough for our purposes.
3. Do not taste sample, even to test for halite.
4. Take care of the mineral samples—no deep scratches, excessive peeling of mica sheets, or breaking samples to determine cleavage.

Mineral Identification

Your teacher may provide you with a laboratory data sheet for you to complete, or you may create your own table in your Lab Journal. Your table should have columns for mineral properties, and rows for sample numbers, as shown below.

Sample No.	Luster	Hardness	Color	Streak	Other Properties	Mineral Name	Chemical Formula	Mineral Group
1								
2								
etc.								

Use the mineral identification guide in Appendix A to identify the mineral samples, according to the following guidelines:

1. It is not necessary to identify the samples in any particular order.
2. The first step is to observe whether the luster is metallic or nonmetallic. Minerals with metallic luster are listed in Part 1 of Appendix A, and minerals with nonmetallic luster in Part 2.
3. After observing the luster, determine the hardness, color, and streak of the sample.
4. Sometimes the most diagnostic mineral property will be in the "other properties" column, such as cleavage for micas, magnetism for magnetite, or reaction with hydrochloric acid.
5. Based on the sample's properties, identify the mineral and write its name in the appropriate column.
6. Use Appendix A to complete the chemical formula column. Examples of mineral groups include silicates and oxides.

Questions

1. Which samples are you confident you properly identified? Why?
2. Are there samples for which you were not as sure about the correct identification? Explain why your identifications were less certain for these samples.

Chapter 5

Rocks and the Rock Cycle

Geologists often determine quite a bit about a rock by examining it closely with a magnifying lens, but they also employ a number of more sophisticated tools to study rocks in depth. One of the most commonly used tools is a specially-equipped microscope for viewing paper-thin slices of rocks. Rocks are sliced with a diamond saw, glued to a microscope slide, and then polished until they are only 30 micrometers (0.03 mm) thick. The resulting microscope slide is called a thin section. *The above photograph is a thin section of a rock called gabbro, composed of olivine, pyroxene, and plagioclase feldspar. The olivine grains are blue and green, the pyroxene grains have a rough gray appearance, and the plagioclase grains have a series of parallel lines in them. Geologists use thin sections to estimate mineral percentages, and to gather clues about the rock's history.*

Objectives

After studying this chapter and completing the exercises, you should be able to do each of the following tasks, using supporting terms and principles as necessary.

1. Draw the rock cycle, with arrows and labels.
2. Describe the difference between magma and lava.
3. Classify and identify igneous rocks.
4. Distinguish between clastic, organic, and chemical sedimentary rocks.
5. Classify and identify sedimentary rocks.
6. Describe how heat and pressure can change rocks.
7. Classify and identify metamorphic rocks.
8. Explain how deposits of coal, petroleum, and natural gas form.
9. Compare and contrast renewable and nonrenewable energy resources.

Vocabulary Terms

You should be able to define or describe each of these terms in a complete sentence or paragraph.

1. biomass
2. bituminous coal
3. cap rock
4. cementation
5. clast
6. compaction
7. composition
8. contact metamorphism
9. extrusive
10. felsic
11. foliation
12. fossil fuel
13. geothermal
14. igneous rock
15. intrusive
16. lignite
17. mafic
18. metamorphic
19. natural gas
20. peat
21. petroleum
22. regional metamorphism
23. reservoir rock
24. sediment
25. sedimentary rock
26. source rock
27. texture
28. weathering

You should be able to describe and identify the following common rocks and sediments.

Igneous Rocks

1. andesite
2. basalt
3. diorite
4. gabbro
5. granite
6. obsidian
7. pegmatite
8. porphyry
9. pumice
10. rhyolite
11. scoria

Sediments and Sedimentary Rocks

12. clay
13. coal
14. conglomerate
15. coquina
16. gravel
17. limestone
18. rock salt
19. sand
20. sandstone
21. shale
22. silt
23. siltstone

Metamorphic Rocks

24. anthracite
25. gneiss
26. marble
27. phyllite
28. quartzite
29. schist
30. slate

5.1 The Rock Cycle

Every rock has a story to tell. The story might be one of a long-extinct volcano, an ancient sandy beach, a reef teeming with marine organisms, or the pressure cooker environment deep in Earth's crust. Geologists use an assortment of clues to determine how rocks were formed, which can lead not only to a better understanding of the history of the world God has placed us in, but to the discovery of new resources.

A rock is a natural, solid mixture of one or more types of mineral grains. The thin slice of rock shown at the beginning of this chapter is composed of obvious mineral grains; in this case olivine, pyroxene, and plagioclase feldspar. In this rock, the individual grains are interlocking, sort of like the pieces of a jigsaw puzzle. The piece of sandstone in Figure 5.1 is composed of grains of mainly one mineral—quartz. The individual quartz grains, which are each less than one millimeter in diameter, are naturally cemented to one another to make a solid rock.

Figure 5.1. This piece of sandstone is composed almost entirely of grains of just one mineral—quartz. Not all sandstones, however, are composed only of quartz.

All rocks can be placed in one of three groups or classes: igneous, sedimentary, and metamorphic. *Igneous*[1] rocks form by the cooling and crystallization of magma or lava. Crystallization may happen rapidly on the surface of Earth or underwater, or it may happen very slowly within the crust or mantle.

Sedimentary rocks form by the compaction and cementation of *sediments*. Sediments are loose mineral grains, which might be formed by the breakdown of pre-existing rocks, or by organisms in water that produce hard shells of various kinds. Examples of sediments include gravel, sand, silt, and clay.

Metamorphic[2] rocks form by the application of heat and pressure on pre-existing rocks. Metamorphism usually occurs thousands of feet beneath Earth's surface, where temperatures and pressures are higher. An example of metamorphism is the transformation of a rock containing clay, which forms in low-temperature settings, into a rock consisting mostly of mica.

It should be clear from these definitions that rocks can change over time. These changes can be summarized by a concept called the rock cycle, which is illustrated by Figure 5.2.

1 "Igneous" comes from a Latin word meaning "fire."

2 "Metamorphic" comes from Greek words meaning "change" and "form."

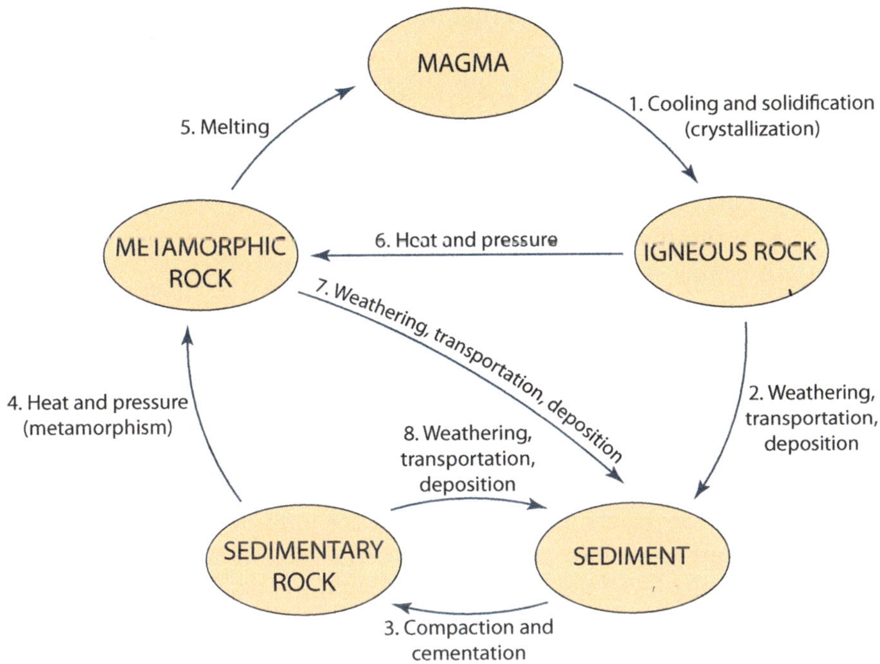

Figure 5.2. The rock cycle.

We can examine the rock cycle starting with magma, at the top of the figure, working our way clockwise around the outside of the diagram. There are five processes around the outside of the rock cycle, represented by arrows, leading to the formation of five different types of substances.

1. When magma cools, it begins to solidify by the formation of crystals. As it cools further, it will eventually come to a point where it is completely solid, forming an igneous rock.
2. If this igneous rock is exposed at Earth's surface, it will be exposed to the day-to-day effects of weather, and will begin to break down to produce loose sediment. This breakdown of rocks to form sediments is called *weathering*. Erosion can transport sediment from one place to another by water, wind, or ice. Eventually, the sediment will be deposited at some place such as an ocean, river floodplain, swamp, or sand dune.
3. If that deposit of sediments becomes buried under additional sediments, it can eventually be compacted and cemented together to form a sedimentary rock.
4. Burial under the weight of thousands of feet of additional sediments will increase the heat and pressure to the point that metamorphism will start to change the rock's minerals, and it will become a metamorphic rock.
5. If the temperature rises far enough, the metamorphic rock will melt, forming magma.

There are a few additional paths that rocks can take as they move through the rock cycle, represented by the arrows on the interior of the rock cycle.

6. Some igneous rocks are never exposed to weathering at Earth's surface. An igneous rock formed near Earth's surface, however, can undergo metamorphism if it is exposed to conditions unlike the conditions at which it formed. Many igneous rocks are at some point squeezed under high pressure, changing them from igneous rocks into metamorphic rocks.
7. Metamorphic rocks, like igneous rocks, can be exposed to the atmosphere and altered into loose sediments.
8. Sedimentary rocks may also be subjected to weathering and disintegrate into their original grains. Note that "sediment" on the rock cycle diagram has three arrows pointing to it, indicating that igneous, sedimentary, and metamorphic rocks can all be weathered and eroded to form sediments.

The rock cycle can pause at one place for very long periods of time. For example, some igneous rocks have never been weathered and eroded into sediment, and never transformed into metamorphic rocks. In other words, they formed some time in Earth history, and have not been significantly altered since then. The rock cycle is a useful concept for understanding the three types of rocks and how they form. In the next three sections, we will take a closer look at igneous, sedimentary, and metamorphic rocks.

Learning Check 5.1

1. Draw the rock cycle, including arrows and labels.
2. Describe the processes of weathering, erosion, transportation, and deposition.
3. Which parts of the rock cycle occur deep in Earth's crust, and which parts occur at or near the surface?

5.2 Igneous Rocks

Most of Earth's crust is composed of igneous rocks, formed by the crystallization of molten rock. Obvious examples of this are the rocks formed by the crystallization of lava from volcanoes such as Mauna Loa in Hawaii or Vesuvius in Italy. Formation of rocks by solidification from molten rock commonly occurs at depth as well, such as in vast magma chambers beneath volcanic areas. Figure 5.3 illustrates the huge magma chamber beneath Yellowstone National Park.

5.2.1 Magma and Lava

Earth's crust is almost entirely solid rock, with only a very small percentage being composed of magma. Rather than being completely liquid, magma is usually a slushy mix of molten rock and solid crystals, as illustrated in Figure 5.4. Magma is composed primarily of the same eight elements that make up 98% of Earth's crust. Additionally, magma contains minor elements with concentrations measured in a few tenths of a percent, such as titanium (Ti) and manganese (Mn), as well as trace elements, which are present in very small concentrations, such as gold (Au), copper (Cu), and uranium (U).

Figure 5.3. There is an immense magma chamber several thousand feet beneath Yellowstone National Park, which hosts one of the world's largest volcanoes.

One might wonder why geologists don't just use the same word for molten rock, whether it is inside the earth or flowing on the surface. Magma inside Earth contains a significant amount of dissolved gases, such as carbon dioxide (CO_2), water (H_2O), and sulfur dioxide (SO_2). As long as the magma remains inside Earth, there is enough pressure from the surrounding rocks to keep the gases dissolved. As soon as the fluid comes out of Earth as lava, such as at a volcano, that pressure is released and the gas comes out of solution, as depicted in Figure 5.5. A useful analogy is a bottle of carbonated beverage. As long as the bottle is tightly capped, there is sufficient pressure inside the bottle to keep the carbon dioxide dissolved in the water. When the cap is removed, gas is released. Figure 5.6 shows an igneous rock that has gas bubbles frozen in place.

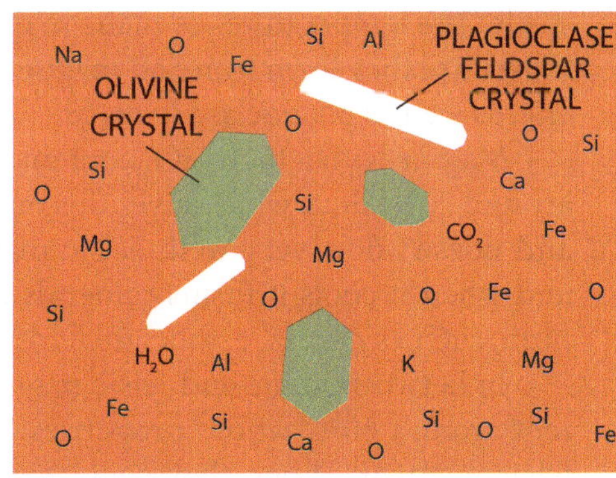

Figure 5.4. Magma is usually a blend of solid crystals in a molten mixture of metal ions and dissolved gases.

Figure 5.5. Gases being released near the summit of a volcano. The visible emissions are formed from water vapor, but the gas also contains carbon dioxide and sulfur dioxide.

5.2.2 Types of Magma and Lava

Figure 5.6. A volcanic rock (scoria) with preserved gas bubbles. The coin is for scale.

Not all magmas or lavas are identical; in fact, there is a wide range of different magma types, each with distinguishing characteristics, and each of which crystallizes to form distinct types of igneous rocks. Here, we will consider just one variable in the composition of magma, which is the amount of silica (SiO_2) in the magma.

The first type of magma to consider is the most common type, which is *mafic* magma. Mafic magma contains a relatively low amount of silica—about 50%; has high concentrations of iron, magnesium, and calcium; and has low concentrations of aluminum, sodium, and potassium. Low-silica magmas tend to flow quite easily. The temperatures of mafic magmas are usually over 1,000°C, and when they crystallize they form iron and magnesium-rich minerals such as olivine and pyroxene, as well as plagioclase feldspar.

Magma with a high SiO_2 content is called *felsic* magma. Felsic magma contains around 70% silica, which gives this magma a high viscosity, meaning that it is sticky and doesn't flow very well. Felsic magma finishes crystallization at about 700°C, and when it cools it forms minerals rich in SiO_2, potassium, sodium, and aluminum, such as quartz, orthoclase feldspar, and muscovite mica.

In between mafic and felsic magma is *intermediate* magma, with a silica content around 60%, whose properties lie somewhere in between that of mafic and felsic magmas. When they cool, these magmas crystallize to form minerals with plagioclase feldspar, amphibole, and pyroxene. These three magma types are summarized in Table 5.1.

As mafic magma or lava cools and crystallizes, the first minerals to form are those rich in iron and magnesium, such as olivine and pyroxene. As the magma cools further, other minerals start to form. Felsic magma, on the other hand, does not contain enough iron and magnesium to form minerals such as olivine and py-

Property	Felsic	Intermediate	Mafic
Approximate temperature	700-800°C	900°C	>1000°C
Silica (SiO_2) content	70%	60%	50%
Concentration of Fe, Mg, Ca	low	medium	high
Concentration of Al, K, Na	high	medium	low
Gas content (H_2O, CO_2, SO_2)	high	medium	low
Concentration of Al, K, Na	high	medium	low
Viscosity (ability to flow)	high viscosity (sticky)	intermediate viscosity	low viscosity (flows easily)

Table 5.1. Basic magma types.

roxene, so the little bit of iron that is there goes into other minerals, such as amphibole and biotite mica. The last mineral to form as felsic magma crystallizes is quartz. The general order in which minerals crystallize, can be summarized as:

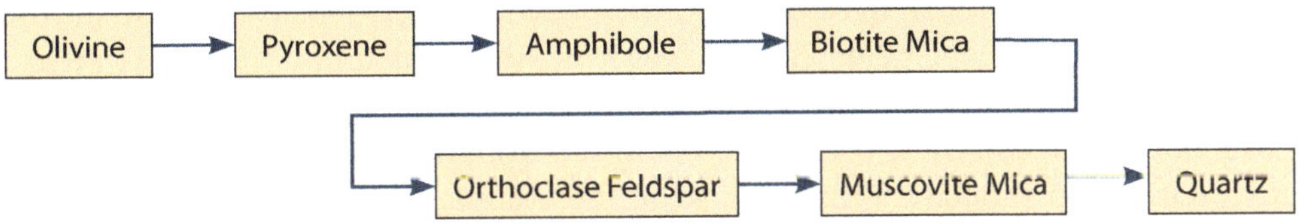

Plagioclase feldspar crystallizes at a range of temperatures, at the same time as olivine, pyroxene, amphibole, and biotite mica. Rocks formed from mafic magma are composed primarily of minerals on the left end of this sequence—olivine and pyroxene—plus plagioclase feldspar. Rocks formed from intermediate magma are composed primarily of pyroxene, amphibole, and plagioclase feldspar. Rocks formed from felsic magma contain minerals from the right end of this mineral sequence: quartz, muscovite mica, orthoclase feldspar, biotite mica, and amphibole.

5.2.3 Intrusive and Extrusive Igneous Rocks

Magma forms deep in Earth's crust, where temperatures are often well above 1,000°C, or even in the upper parts of Earth's mantle. Because liquid rock has a lower density than solid rock, it is buoyant, like a cork in water, and magma tends to rise towards Earth's surface. As it rises, it encounters cooler conditions and begins to crystallize. If it completely crystallizes beneath the surface of Earth, the resulting rocks are called *intrusive* igneous rocks. Intrusive rocks might form just a few hundred meters beneath Earth's surface, but usually they form at depths of several kilometers or greater. Magma deep in the crust will cool and crystallize slowly, so individual crystals have time to grow. The grain sizes in intrusive igneous rock are typically several millimeters across, as in the granite sample shown in Figure 5.7.

Figure 5.7. Granite is an intrusive rock, and has crystals that are visible to the unaided eye.

When lava cools on Earth's surface or under water, it forms *extrusive* igneous rocks. Rather than taking hundreds, or even many thousands of years to crystallize, crystallization occurs in a matter of minutes, hours, or days. The crystals in extrusive igneous rocks do not have time to grow very large, and are for the most part microscopic.

5.2.4 Identifying Igneous Rocks

Texture refers to the size of crystals in the rock. Intrusive rocks have a course-grained texture, meaning that the mineral grains are large enough to be seen without a magnifying glass or microscope. Extrusive rocks, on the other hand, are fine-grained, and most of their crystals are too small to see with the unaided eye.

Composition refers to the chemical elements and minerals a rock is made of. As already mentioned, rocks that form by the crystallization of felsic melts have a lot of light-colored minerals, such as quartz, orthoclase feldspar, and muscovite mica. Not only are these rocks light-colored, they have a lower density because of low concentrations of heavier elements such as iron. The small amount of iron in these rocks goes into the specks of dark-colored minerals, such as hornblende and biotite mica, often visible in these rocks. Rocks formed from mafic magma do not contain any quartz or orthoclase feldspar. Instead, they contain a mixture of plagioclase feldspar along with darker, denser minerals such as pyroxene and olivine, resulting in an overall dark appearance for these rocks.

Many igneous rocks can be named using the simple table shown in Table 5.2. The first step in identifying an igneous rock is to determine its texture. If the minerals are mostly course-grained, the rock is probably granite, diorite, or gabbro. If the minerals are mostly fine-grained, the rock is likely to be rhyolite, andesite, or basalt.

The second step in naming an igneous rock is to determine whether it is felsic, intermediate, or mafic. For course-grained rocks, this is a matter of determining which minerals are present in the rock. In general, mafic rocks will have a darker overall appearance, and felsic rocks will have a lighter overall appearance. A more

	Felsic	Intermediate	Mafic
Course-grained rocks	*granite* • overall light color • minerals: quartz, orthoclase feldspar, biotite mica, muscovite mica, amphibole	*diorite* • overall intermediate color • minerals: pyroxene, amphibole, plagioclase feldspar	*gabbro* • overall dark color • minerals: olivine, pyroxene, plagioclase feldspar
Fine-grained rocks	*rhyolite* light-colored, fine-grained rock	*andesite* gray, fine-grained rock	*basalt* dark-colored, fine-grained rock
Other igneous rocks	*obsidian*—a dark, glassy igneous rock *pumice*—an igneous rock with so many gas bubbles that it can float on water *pegmatite*—a course-grained igneous rock with crystals larger than about 2.5 cm (one inch) *porphyry*—an igneous rock with two distinct sets of crystal sizes *scoria*—a fine-grained igneous rock with numerous gas bubbles (but not enough bubbles to make the rock float)		

Table 5.2. Igneous rock classification.

Mineral	Description
Olivine	usually small, medium-green, non-elongated grains
Pyroxene	dark green stubby crystals
Amphibole	dark, elongated crystals
Biotite Mica	small black flakes
Plagioclase Feldspar	white to gray or black grains; may be somewhat elongated and rectangular; can have fine, parallel lines within crystal faces
Orthoclase Feldspar	pink or white grains
Quartz	gray to transparent glassy grains

Table 5.3. Minerals in coarse-grained igneous rocks.

accurate identification of a rock can be made by actually identifying the minerals that are present in the rock. Table 5.3 can serve as a guide to identifying visible minerals in igneous rocks.

Due to small crystal size, it is not possible to identify individual mineral grains in a fine-grained rock without a microscope. There are two ways to get around this. The first is overall color of the rock. If it is lighter colored, it is probably rhyolite, shown in Figure 5.8; if it is almost black it is likely to be basalt. The second clue to a fine-grained rock's name is that there are often at least a few crystals that are large enough to identify. If the rock contains visible crystals of olivine, it must be basalt. If a fine-grained rock contains crystals of plagioclase feldspar, it is either andesite or basalt.

Figure 5.8. Volcanic rocks are extrusive rocks with mostly microscopic crystals. This light-colored rock is rhyolite.

There are five additional rock names at the bottom of Table 5.2. *Obsidian*, shown in Figure 5.9, is an igneous rock formed when lava cools extremely rapidly, such as when lava comes in contact with water or ice. When this happens, there is no time for crystals to form, even at a microscopic level, resulting in the formation of a volcanic glass rather than crystals. As shown back in Figure 4.9, atoms in glass are

Figure 5.9. Obsidian, pumice, and pegmatite.

arranged randomly in the solid rather than being in an orderly pattern. Obsidian is typically black, and can break into razor-sharp fragments. In the past, obsidian was often used for making stone tools such as arrowheads; now it is sometimes used to make ultra-sharp surgical scalpels.

Pumice is also a rock that is made predominantly of volcanic glass, but is formed when lava is thrown up into the atmosphere above a volcano. Pumice, shown in Figure 5.9, contains a multitude of tiny gas bubbles. This makes the overall rock less dense than water and enables pumice to float on water. *Scoria* is like pumice, containing a large number of gas bubbles, but not enough bubbles to allow the rock to float on water.

Pegmatite, also shown in Figure 5.9, is an intrusive igneous rock, in many cases of granitic composition, that has very large crystals. The crystals in pegmatites are typically about 5 centimeters across, though in extreme cases, crystals several meters long have been found.

Figure 5.10. Andesite porphyry with large crystals of plagioclase.

In many cases, magma will start to solidify at some depth beneath a volcano, forming coarse-grained crystals. When a partially-crystallized lava erupts out of a volcano, the remaining fluid will crystallize rapidly, resulting in a rock that is a combination of intrusive and extrusive. This gives the rock both coarse and fine mineral grains, as in the rock shown in Figure 5.10. The resulting rock is called a *porphyry*. Generally, a porphyry is named as a type of fine-grained rock, such as rhyolite porphyry or basalt porphyry.

Learning Check 5.2

1. Compare and contrast intrusive and extrusive igneous rocks.
2. Distinguish between magma and lava.
3. Why does granite have larger crystals than rhyolite?
4. What two properties are used to classify igneous rocks?

5.3 Sedimentary Rocks

Sedimentary rocks are formed when loose sediments are compacted and cemented together to make a solid rock. Many sedimentary rocks, such as sandstone, are composed of fragments of pre-existing rocks. Other sedimentary rocks are made primarily of the skeletons of once-living organisms. Limestone is composed mainly of calcium carbonate ($CaCO_3$) fragments formed by a variety of marine or freshwater organisms. Still other sedimentary rocks form by processes such as evaporation of seawater.

5.3.1 Clastic Sedimentary Rocks

If rocks are exposed at Earth's surface, they will begin to experience weathering, a combination of chemical and physical processes that loosen and change the grains in a rock, eventually producing soil. Some minerals in rocks, such as quartz, are very resistant to weathering. Other minerals, such as the feldspars and micas, are more vulnerable to weathering, and transform into minerals such as various types of clay. Over time, a solid piece of granite can be transformed into a sandy mixture of unaltered quartz, partially-weathered feldspar, and clay, as illustrated back in Figure 4.16.

Once loosened by weathering, mineral grains are free to move downhill through the process of *erosion*—the removal of loose sediments by wind, water, or ice, or gravity. Figure 3.15 illustrated soil erosion by water. Once eroded from a hillside, sediments can be transported long distances. Sand grains can roll or bounce along the bottom of a stream. A sand grain eroded from the top of the Rocky Mountains can eventually be transported thousands of miles and end up on a beach or on the sea floor in the Pacific Ocean or Gulf of Mexico. Deposition occurs when sediments stop being transported and are laid down in a place such as a river mouth or beach, and then get buried under additional sediments.

Loose sediments usually have a large amount of space between the individual mineral grains. As a deposit of sediments gets buried deeper under additional sediments, pressure will cause the sediments to undergo *compaction*, reducing the amount of space between grains. Water in the spaces between grains contains dissolved minerals, which can crystallize to form cement that holds mineral grains together, a process known as *cementation*. Compaction and cementation are illustrated in Figure 5.11. The most common cements in sedimentary rocks are calcite and silica. The combination of compaction and cementation acts to transform loose sediments into a solid sedimentary rock.

A *clastic* sedimentary rock is one that is formed of clasts—mineral or rock fragments—eroded from previously existing rocks. These clasts might be individual mineral grains, such as pieces of quartz eroded from granite, or they might be larger fragments made up of multiple minerals. Clastic sediments and sedimentary rocks are classified primarily by grain size, as outlined in Table 5.4.

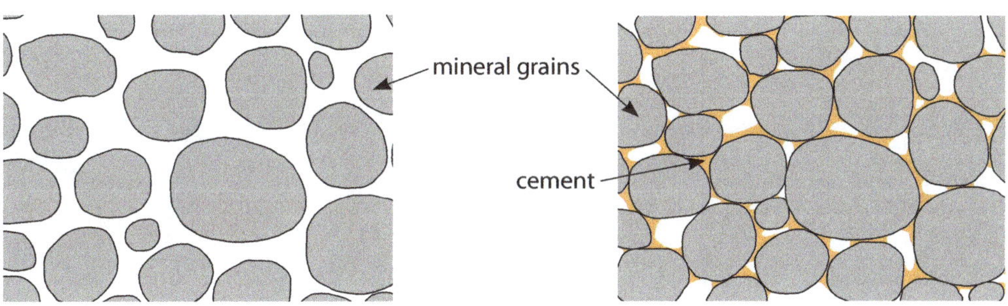

Figure 5.11. Compaction and cementation.

Sediment Name	Rock Name	Grain Size
gravel	conglomerate	> 2 mm
sand	sandstone	0.06 mm–2 mm
silt	siltstone	0.004 mm–0.06 mm
clay	shale	< 0.004 mm

Table 5.4. Clastic sediments and sedimentary rocks.

Conglomerate is composed of grains of gravel that are larger than 2 mm, though the grains can be considerably larger than this. Gravel is most typically deposited near hilly or mountainous areas, though large streams can transport gravel for long distances.

Sand is deposited in a variety of environments, such as streams, stream deltas, lakes, beaches, and sand dunes. When sand grains become cemented together, the resulting rock is *sandstone.* Many sandstones are composed largely of grains of pure quartz. The minimum size for sand grains is 0.06 mm, which is about the size of the smallest grains visible to the human eye without a magnifying glass. If the grains are too small to be seen with the unaided eye, then the rock is likely to be a *siltstone* or *shale.* Siltstone is a sedimentary rock composed of microscopic mineral grains that are larger than 0.004 mm but smaller than the limit of human vision, 0.06 mm. The mineral grains in siltstone are often primarily quartz. Shale, on the other hand, is composed of extremely tiny mineral grains that cannot be studied in detail even with a good standard microscope. Rather than being composed mostly of quartz fragments, shale is usually composed of clay. Because of their microscopic grain size, it can be difficult to distinguish shale and siltstone. Often it is possible to see the individual grains in siltstone with a good magnifying glass. Shale often has distinct, thin layers, while siltstone often lacks these layers. In addition, the surface of siltstone can feel gritty when rubbed across the fingernails. Conglomerate, sandstone, and shale are illustrated in Figure 5.12.

5.3.2 Organic and Chemical Sedimentary Rocks

Not all sedimentary rocks form by the cementing together of clastic grains. *Limestone* is a rock that is composed mostly of calcite ($CaCO_3$). The calcite grains that make up limestone are mostly formed from the remains of marine or freshwater organisms such as algae, clams, snails, and corals. These organisms take $CaCO_3$

Figure 5.12. Clastic sedimentary rocks. Siltstone is not illustrated, but often looks similar to sandstone, except the grains are too small to be seen.

Figure 5.13. Limestone varieties.

that is dissolved in the water and use it to form hard shells or skeletons. When these organisms die, their soft parts decay or are eaten by other organisms, but the solid $CaCO_3$ particles remain. These skeletal remains often break up into smaller fragments. The grains produced by algae can be of microscopic size, and actually make up the bulk of most limestones. When the fossil grains become cemented together, the resulting rock is limestone.

Not all limestones are identical to each other. Some are made entirely of microscopic grains of calcite; for now it will be good enough to just call these "limestone." Many limestones contain numerous fossils, and these are called *fossiliferous limestone*. A fossiliferous limestone that is composed almost entirely of fossil shell fragments is known as *coquina*. These three types of limestone are shown in Figure 5.13.

Coal is an organic sedimentary rock[3] formed by the compaction and hardening of decayed plant material. We know this because of the fossilized fern leaves that occur in coal, such as the one shown in Figure 5.14. Most coal seems to have formed from the decay of vegetation that grew in prehistoric swamps. Trees and other plants died and sank to the bottom of the swamps. As more vegetation lived and died, the layers of decayed organic material became thicker and thicker. Eventually the plant material is transformed into coal by heat, pressure and chemical reactions.

Figure 5.14. Coal containing a fossil fern leaf.

In some cases, rocks can be formed directly from minerals dissolved in seawater. As illustrated by the salt evaporation pond back in Figure 4.12, halite and other minerals can form by evaporation of seawater. When this happens in nature, very thick deposits of minerals such as halite and gypsum can form. When halite forms a rock, it is often called *rock salt*.

---

3 Even though "rock" is defined in Section 5.1 as a mixture grains, and coal is not considered to be a mineral because it is organic, geologists commonly consider coal to be a rock.

5.3.3 Identifying Sedimentary Rocks

Table 5.5 summarizes the various types of sedimentary rocks. The most important distinction among sedimentary rocks is between clastic rocks and non-clastic rocks, such as limestone, coal, and rock salt.

Conglomerate and sandstone are generally easy to identify, since you can see the individual sand or gravel grains in the rock. Limestone is typically easy to identify as well, because it reacts vigorously with dilute hydrochloric acid. If their grains are cemented together by calcite, clastic rocks can react with hydrochloric acid as well, but the fizzing is much less vigorous.

Siltstone and shale are distinguished from each other by the thin layers present in shale. These layers range in thickness from paper-thin to a millimeter or two. Shale occurs in a variety of colors, including tan, red, gray, and black.

Coal is usually black or very dark brown, and may contain black plant fossils. Other sedimentary rocks are dark-colored as well if they contain a large amount of organic material, especially limestone or shale. Limestone can be distinguished

Clastic sedimentary rocks—derived from weathering of pre-existing rocks			
Texture	**Grain Size**	**Composition**	**Rock Name**
clastic (rock and mineral fragments)	gravel (pebbles, cobbles, and/or boulders) > 2 mm	mostly quartz, feldspar, clay, and micah; with some rock fragments and other minerals	conglomerate
	sand 0.06 to 2 mm		sandstone
	silt 0.004 to 0.06 mm		siltstone
	clay < 0.004 mm	mostly clay minerals	shale
Chemical and organic sedimentary rocks			
Composition	**Texture**	**Comments**	**Rock Name**
calcite ($CaCO_3$); reacts with dilute hydrochloric acid	microscopic crystals	no or few fossils	limestone
	mix of microscopic crystals and fossils	contains fossils	fossiliferous limestone
	fossils	mostly fossils	coquina
halite (NaCl)	crystalline		rock salt
organic material	microscopic to coarse-grained	black or dark brown	coal

Table 5.5. Sedimentary rock classification.

from coal by the hydrochloric acid test, and coal does not usually have the thin layers that are common in shale.

Learning Check 5.3

1. Explain how loose sediments can be changed into solid sedimentary rocks.
2. What are the four sizes of clastic sediments, and what are the corresponding types of sedimentary rocks?
3. Why might a quartz sandstone slightly fizz when a drop of hydrochloric acid is placed on it?
4. Compare and contrast the grains found in limestone with those found in clastic rocks.

5.4 Metamorphic Rocks

Metamorphic rocks can form from other rocks through heat, pressure, or a combination of the two. Metamorphism can be mild, making only minor changes to the appearance of a rock, or it can be extreme, transforming one kind of rock into something that looks entirely different. An important thing to remember is that metamorphic changes occur in the solid state. If the temperature becomes hot enough to melt minerals in the rock, the result is an igneous process, not a metamorphic one.

5.4.1 Contact Metamorphism and Regional Metamorphism

I have already mentioned that temperature increases with depth in Earth's crust. Another place where it gets very hot in Earth's crust is adjacent to magma chambers beneath volcanoes, such as shown in Figure 5.3. Here, the high temperature of the magma can "bake" surrounding rocks, metamorphosing minerals from one kind to another. This high-temperature metamorphism due to the presence of magma is called *contact metamorphism*. Contact metamorphism often occurs in a zone close to intrusive igneous rocks, as illustrated in Figure 5.15. In some cases, contact metamorphism can cause only minor changes to the cooked rocks. In other cases, especially when the rock being metamorphosed is limestone, the changes can be dramatic.

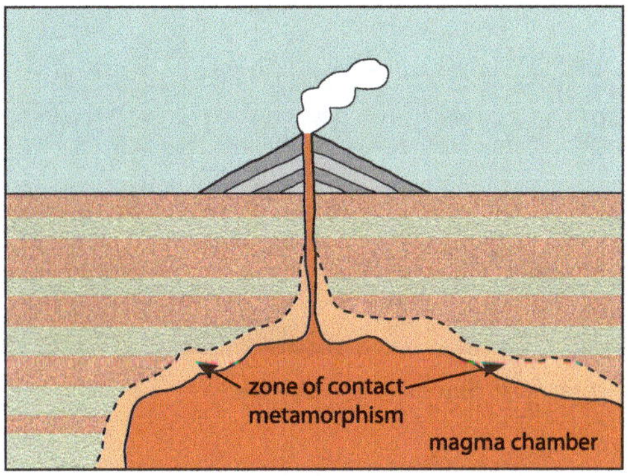

Figure 5.15. The zone of contact metamorphism around an underground body of magma.

The increase in pressure with depth within Earth's crust is similar to what you experience if you go down to the bottom of the deep end in a swimming pool, where you can feel the pressure from the weight of the water above you pushing on your ear drums. Deep in the earth, the pres-

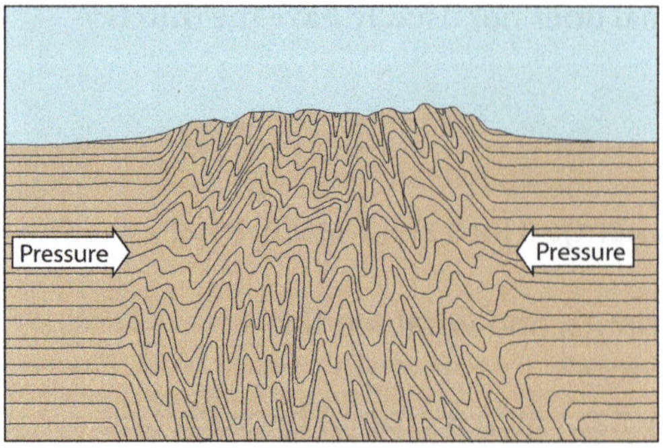

Figure 5.16. Regional metamorphism occurs when rocks are compressed with lateral (sideways) pressure during the uplift of mountain ranges.

sure caused by the weight of overlying rocks is many thousands or even millions of pounds per square inch. This pressure causes mineral grains to rearrange within a rock, and even causes atoms to rearrange within crystals to form new minerals. Another important type of pressure that causes metamorphism is lateral, or sideways, pressure. This lateral pressure often occurs deep in the crust beneath mountain ranges as they form, and is called *regional metamorphism*, illustrated in Figure 5.16. Contact metamorphism only occurs close to magma in Earth's crust, but areas of regional metamorphism can extend for hundreds of kilometers in the roots of mountain ranges.

5.4.2 Identifying Metamorphic Rocks

One way to illustrate the process of metamorphism is to look at the changes that happen to the sedimentary rock shale as it undergoes regional metamorphism. These changes start out as minor, but with increasing metamorphism, shale is transformed into a radically different rock. The following sequence summarizes the metamorphism of shale over time:

Shale → Slate → Phyllite → Schist → Gneiss

There are two types of changes that occur in this process. The first is a change of minerals in the rock. Shale is composed mainly of clay, a low-temperature mineral. As temperature and pressure increase, the clay becomes unstable and changes to mica, a higher-temperature mineral. Slightly metamorphosed shale is known as *slate*. Slate looks similar to shale, but is generally harder. Like the clay in shale, the mica grains in slate are microscopic and slate has a dull appearance. Slate often splits into very flat sheets, and is used to make chalk boards, pool table surfaces, and roofing shingles, as shown in Figure 5.17. As metamorphism progresses, slate changes into *phyllite*, in which the mica grains are larger but still microscopic. Because the mica flakes are larger, phyllite has a glossy appearance. Further metamorphism will yield *schist*, in which the mica grains are large enough to be clearly seen. The most advanced stage of regional metamorphism is *gneiss*. In gneiss (pronounced "nice"),

Figure 5.17. Slate roofing shingles.

Figure 5.18. Phyllite, schist, and gneiss.

the micas are transformed into quartz, feldspar, and amphibole. Gneiss can look very much like granite, but often has distinct bands of light- and dark-colored minerals. Phyllite, schist, and gneiss are shown in Figure 5.18.

A common feature of metamorphism is *foliation*—which is the alignment of minerals to create new layers. Illustrated in Figure 5.19, foliation is caused by intense lateral pressure, which makes flat minerals such as mica line up so mineral grains are parallel to each other. When shale is metamorphosed into slate, the mica grains line up with each other, making slate a foliated rock. Slate, phyllite, schist, and gneiss are all foliated rocks.

I have described what happens to shale as it is metamorphosed. Other rocks will go through different metamorphic paths, often producing different end products. Many sandstones, for instance, are composed almost entirely of quartz (SiO_2). Because there is little aluminum, potassium, magnesium, or iron in quartz sandstone, it is impossible for pure quartz to produce mica or any other mineral. Typically the only thing that happens to a quartz sandstone when it is metamorphosed is that the grains get squished together, eliminating the pore space between sand grains. The resulting rock is *quartzite*. The same sort of thing happens to limestone. Metamorphism will cause the calcite grains in limestone to grow larger and rearrange to form *marble*. Coal, when it is metamorphosed, becomes a hard, black, shiny variety called *anthracite*.

Table 5.6 summarizes some important metamorphic rocks.

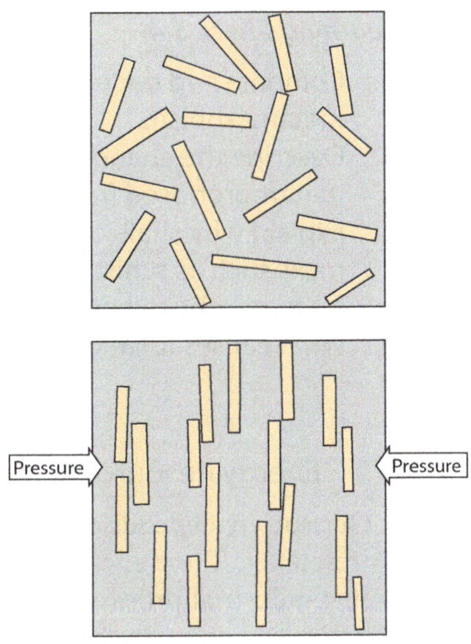

Figure 5.19. Foliation is produced in metamorphic rocks when pressure forces elongated minerals to rotate so they are parallel to each other.

	Rock Name	Grain Size	Minerals	Comment	Parent Rock
Foliated	slate	microscopic	mica	breaks in flat sheets with dull surfaces	shale
	phyllite	microscopic	mica	breaks along wavy surfaces; glossy appearance	shale
	schist	medium	mica + others	micas are dominant	shale, granite, volcanic rocks
	gneiss	medium to coarse	feldspar, quartz, amphibole + others	similar to granite, but dark and light minerals are usually in distinct bands	shale, granite, volcanic rocks
Non-foliated	quartzite	fine to medium	quartz	hard, light-colored rock made of quartz	sandstone
	marble	medium to coarse	calcite	interlocking calcite grains; hydrochloric acid causes fizzing	limestone
	anthracite	fine	organic	hard, black, shiny rock	coal

Table 5.6. Metamorphic rock classification.

Learning Check 5.4

1. Compare and contrast regional metamorphism and contact metamorphism.
2. Describe the changes in texture and minerals as shale is progressively metamorphosed from slate to gneiss.
3. Explain why shale can be metamorphosed into very different looking rocks such as schist and gneiss, but quartz sandstone can only be metamorphosed into quartzite.
4. What causes foliation in metamorphic rocks?

5.5 Energy Resources

Our society depends on the Earth not only for mineral resources, but for energy resources. These energy resources are used as fuels for generating electricity, running industrial processes, and powering transportation. Some of these energy resources are nonrenewable, such as coal and petroleum. Earth's crust contains a limited amount of these resources, so they could one day be in short supply. Other energy resources are renewable, such as wind, solar, geothermal, and hydroelectric power.

5.5.1 Coal

The main fuels used in our modern society are coal, petroleum, and natural gas. These three substances are known as *fossil fuels* because they all originated from the decay of organisms that lived on Earth in the distant past.

Figure 5.20. Coal being mined at a surface mine in Wyoming.

Coal is a solid fossil fuel composed of compressed and altered plant remains. As the plant material undergoes heat and pressure, it is compacted and chemical reactions occur that transform the plant material into *peat*. Peat is a spongy material containing partially-transformed pieces of plants. Today, peat is found in large deposits in places like Scotland and Russia, where it is scraped off Earth's surface and used as a low-energy fuel. Peat is not considered to be coal.

Upon further compaction, peat changes into *lignite*, a soft brown variety of coal. Lignite produces a moderate amount of heat when it is burned, though it generates more heat than the same amount of peat. Further transformation leads to the formation of *bituminous coal*, which is usually black. Bituminous coal gives

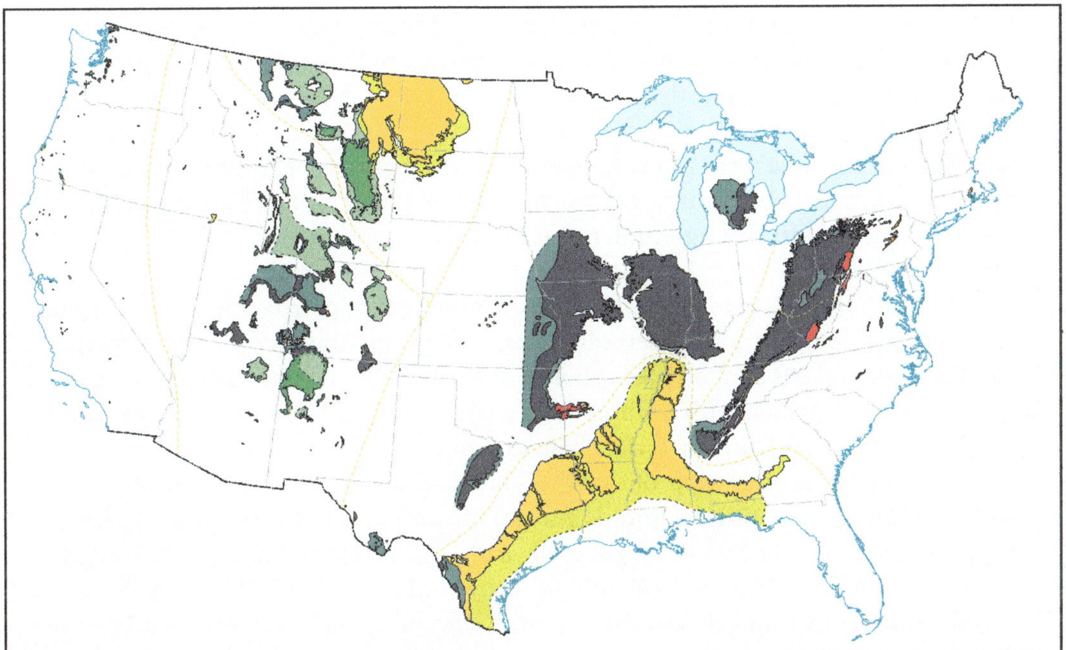

Figure 5.21. The distribution of major coal deposits in the United States. Yellow regions are lignite; tiny orange regions in Pennsylvania are anthracite; others are various grades of bituminous. Darker colors indicate areas of commercial value.

off a high amount of heat when burned. If bituminous coal is metamorphosed, it becomes anthracite, a hard, black, shiny variety of coal that produces a higher amount of heat when burned than all other grades of coal. However, anthracite is not very common and bituminous coal is the type most commonly used as a fuel for generating electricity.

The following steps summarize the formation of coal:

Plants → Peat → Lignite → Bituminous Coal → Anthracite

Fossils and other sedimentary rocks associated with coal indicate that coal forms from vegetation that grew in swampy environments. Both laboratory and field evidence indicate that the transformation of this organic material into coal requires burial under additional sediments and heating up to about 100–150°C.

Figure 5.20 shows a thick layer of coal being mined in the western United States, and Figure 5.21 shows where coal layers are found in the United States. At current rates of mining, there is enough coal in the United States to last for a few hundred years. Because peat and coal take a long time to form, they are not considered to be renewable resources.

5.5.2 Petroleum and Natural Gas

Petroleum[4], often referred to as *crude oil*, or even just *oil*, is a liquid fossil fuel composed of a wide variety of hydrocarbon molecules, molecules composed of carbon and hydrogen. Petroleum is pumped out of the ground and separated into various types of compounds. These are further modified at oil refineries to produce a variety of fuels, such as gasoline and kerosene, as well as a number of products ranging from asphalt to plastics.

Natural gas is a fossil fuel composed primarily of methane, a hydrocarbon with the chemical formula CH_4. Many homes use natural gas as fuel for furnaces, water heaters, and gas stoves. Natural gas has the advantages of being relatively plentiful and easy to transport through underground pipes leading to individual homes and other facilities.

There are three requirements for the accumulation of petroleum or natural gas underground: source rocks, reservoir rocks, and a trap. The *source rock* for petroleum and natural gas is often black shale, containing a high content of decayed organic matter, deposited with sediments in ancient oceans. With time and elevated temperatures, this organic material changes into petroleum and natural gas.

These hydrocarbons seep out of the shale and accumulate in *reservoir rocks*, which hold large quantities of liquids and gas in tiny pores between mineral grains. Layers of sandstone or limestone generally act as the reservoir rock for oil and gas.

Once in a reservoir rock, oil and gas move either horizontally along the reservoir rock layer or upwards towards Earth's surface. If the reservoir rock layer is

4 The word petroleum comes from the Greek word *petra*, which means "rock," and the Latin word *oleum*, which means "oil."

connected to Earth's surface, or if the rocks above the reservoir allow for movement of fluids, the petroleum and natural gas will leak out at the surface and be lost. However, if the once-horizontal rock layers are somehow contorted in the subsurface, the oil and gas may be trapped. An essential component of a trap is the presence of a *cap rock*—a layer that prevents liquids and gases from escaping from the reservoir. Two common types of petroleum traps are illustrated in Figure 5.22. In the first diagram, oil and gas are trapped beneath the cap rock in a *folded* rock structure. In the second, the oil and gas are trapped against a *salt dome*—a pillar of salt that has pushed up from deeper rock salt layers. Note that in each of these settings, the natural gas, being of lowest density, is above the petroleum.

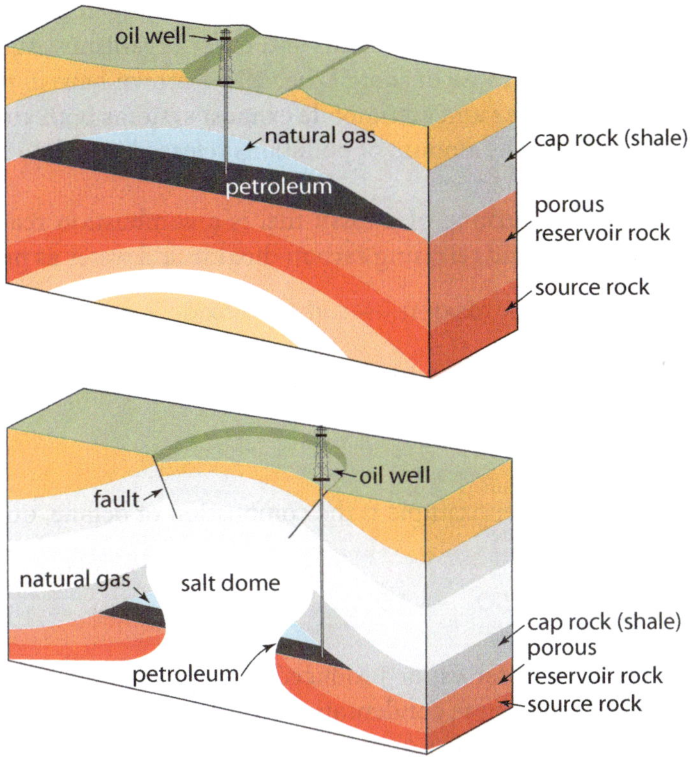

Figure 5.22. Two examples of petroleum traps.

5.5.3 Environmental Concerns with Fossil Fuels

Oil, coal, and natural gas are used to provide much of the world's electricity and most of the world's fuel for transportation. These are, however, limited, nonrenewable resources. Most of the fossil fuels we use are those that are most accessible or easy to find. As we deplete these "easy" resources, we will be increasingly dependent on finding and extracting supplies that are deeper, of poorer quality, or in more remote places. Eventually our demand for these resources will exceed the supply, and we will be forced either to use less energy or to find alternative ways of producing electricity and fueling our transportation.

The burning of fossil fuels, especially coal and oil, produces compounds that cause air pollution. Coal usually contains some pyrite (FeS_2), and when coal is burned, the sulfur is released to the atmosphere as sulfur dioxide (SO_2), leading to the formation of sulfuric acid when the SO_2 combines with atmospheric moisture. This sulfuric acid makes rain and snow more acidic than they naturally would be. This *acid rain* can cause damage to lakes, soil, and plants in a large region downstream from coal-burning power plants. The burning of gasoline in cars releases hydrocarbons, carbon monoxide, and nitrogen oxides, all of which can be harmful to the health of humans and other organisms. Automobile exhaust is the most important cause of smog,[5] the yellow-brown haze in the air over many cities. Coal smokestacks and automobile exhaust systems both contain expensive components to reduce the amount of pollutants released into the atmosphere, but these are only partially effective.

The burning of a fossil fuel is a combustion reaction, requiring oxygen as a reactant, and releasing carbon dioxide and water as products:

$$\text{fossil fuel} + \text{oxygen} \rightarrow \text{carbon dioxide} + \text{water}$$

An example of this chemical reaction is the combustion of methane:

$$CH_4 + 2O_2 \rightarrow CO_2 + 2H_2O$$

Another example is the combustion of octane, one of the main compounds in gasoline:

$$2C_8H_{18} + 25O_2 \rightarrow 16CO_2 + 18H_2O$$

The water vapor produced through combustion of fossil fuels is visible on a cold morning as condensation (often mistakenly called "steam") coming out of automobile tail pipes or power plant smokestacks. Carbon dioxide is an invisible gas, but it is also a greenhouse gas, meaning that it contributes to keeping Earth's atmosphere warmer than it otherwise would be. Most atmospheric scientists believe that the addition of too much carbon dioxide to the atmosphere will cause temperatures to increase, a situation commonly referred to as "global warming." We will take a closer look at greenhouse gases and carbon dioxide in Chapter 16.

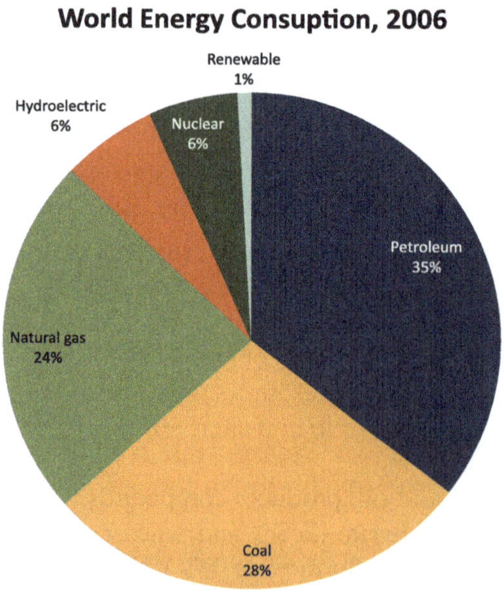

Figure 5.23. World-wide energy use in 2006.

Human beings are using resources such as coal and petroleum as if they will never run out, and yet there is clearly a limited amount of these fuels we can extract from Earth's crust. This brings us back to the issue of stewardship over the creation. How do we balance our need for energy with the fact

5 "Smog" is a combination of the words smoke and fog, though modern smog doesn't require either smoke or fog to form.

that fossil fuels are a nonrenewable resource? How much pollution from the burning of fossil fuels is acceptable, and how can we minimize the impact of this pollution both on human communities and the natural world? There is a great need for Christians to think about such issues.

Figure 5.24. As long as the Columbia River flows, power plants at dams such as the Grand Coulee Dam will be able to generate electricity.

5.5.4 Alternative Energy Resources

As illustrated in Figure 5.23, our society is highly dependent on fossil fuels. About 87% of the energy used in the world comes from nonrenewable fossil fuels. Most of the remaining energy comes from hydroelectric and nuclear power. Hydroelectric power is generated at dams and other facilities along streams, such as the dam shown in Figure 5.24. Hydroelectric power, which supplies about 6% of the world's energy consumption, is considered to be a renewable energy source. Nuclear power also provides about 6% of the world's energy needs. Nuclear power plants, such as shown in Figure 5.25, are powered by uranium, a nonrenewable metal mined from Earth's crust.

The remaining 1% of human energy use comes from other renewable resources. The most important of these renewable energy resources are:

- *Wind power*—At one time, windmills dotted the prairies of the United States. These windmills were used to pump groundwater to the surface for crop irrigation and livestock. Today, there is a renewed push to build giant windmills for generating electricity. Wind is certainly a clean and renewable energy source, but there are disadvantages as well. When the wind is

5.25. Nuclear power plants are dependent on uranium, which must be mined from Earth's crust.

not blowing sufficiently, wind-powered turbines cannot generate electricity. The towers are considered unattractive by many, and have even been known to kill migratory birds.

- *Solar power*—Energy from the sun can be used to heat homes and water, and to generate electricity. There are few downsides to solar power, other than that it doesn't work well under cloudy skies, and doesn't work at all at night time.
- *Geothermal*—Geothermal energy is generated by heat within Earth's crust. Electricity can be generated by running steam from deep in Earth through electrical turbines, or heat can be used directly to heat homes. Geothermal energy is only a practical resource in areas where there is a higher than normal amount of heat in Earth's crust, such as in volcanic areas.
- *Biomass*—Biomass refers to plants that can be used as fuel or that can be converted into fuel. People have been using biological materials, such as wood and dried animal dung, as their primary source of energy throughout human history. Plants can also be grown specifically for the purpose of being burned in power plants. For example, in Hawaii, close to 10% of the state's electricity is generated by burning by-products from sugar cane plants. A significant amount of corn grown in the United States is converted into ethanol, which is then mixed with gasoline and used as fuel for automobiles.

These renewable energy resources will all be increasingly important parts of our energy future. It is unlikely, however, that any of these would be able to provide as much electricity and fuel as are currently provided by fossil fuels. Developing energy resources will be one of the great challenges facing the human race in the upcoming century.

Learning Check 5.5

1. Compare and contrast coal, petroleum, and natural gas.
2. What are advantages and disadvantages of using fossil fuels as energy resources?
3. Describe how petroleum and natural gas deposits are formed.
4. Describe some advantages and disadvantages of developing renewable energy resources.

Chapter 5 Exercises

Answer each of the questions below as completely as you can. Write your responses in complete sentences unless instructed otherwise.

1. "Rock Thin Section 1" depicts a microscopic view of a thin slice of a coarse-grained igneous rock.
   a. Devise a method to estimate the percentage of each mineral in the

rock.

b. Make a data table, listing the minerals and their estimated percentages.

c. Determine the name of this igneous rock.

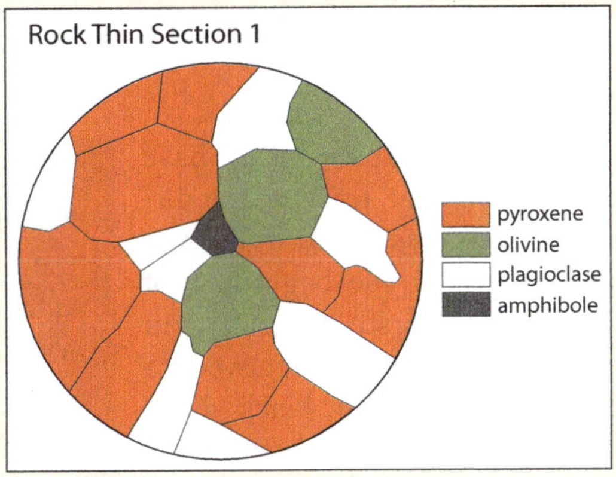

2. "Rock Thin Section 2" depicts a microscopic view of a thin slice of a coarse-grained igneous rock.
   a. Devise a method to estimate the percentage of each mineral in the rock.
   b. Make a data table, listing the minerals and their estimated percentages.
   c. Determine the name of this igneous rock.
3. What are the three main compositional types of magma, and how do they differ from one another?
4. Why are quartz grains found more commonly in sandstone than grains of feldspar, amphibole, or mica?

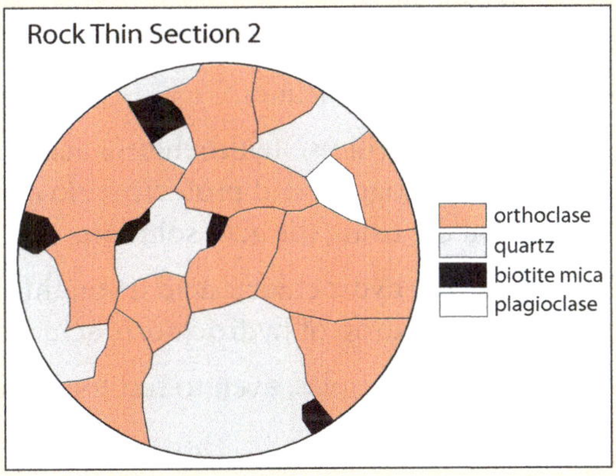

5. Where do marine organisms get the calcium carbonate they need to build shells?
6. How can a sedimentary rock become a different sedimentary rock without first becoming a metamorphic rock?
7. Which type of magma would you expect to cause the greatest amount of contact metamorphism—felsic magma or mafic magma?
8. Why does the rock cycle diagram not have an arrow going directly from sedimentary rock to magma?
9. Metamorphism occurs in the solid state, so metamorphism often takes many thousands of years to occur. If magnesium ions can travel 0.0001 mm/year through olivine, how many years would it take a magnesium ion to travel 0.5 mm in an olivine crystal?

Experimental Investigation 3: Identifying Rocks

Overview

In this chapter, you have learned about at least thirty different rock types, listed in the vocabulary list at the beginning of Chapter 5. At first, it might seem difficult to distinguish between a piece of black basalt, black shale, and black slate, but by carefully examining rock samples, you should quickly become proficient at identifying rocks.

Geologists use properties you have studied in Chapter 5, such as grain size, color, mineral content, and layering, to identify a wide variety of igneous, sedimentary, and metamorphic rocks.

Basic Materials List

- Rock samples
- Dilute hydrochloric acid in small dropper bottle
- Magnifying glass
- Safety goggles and protective gloves

Safety and Other Guidelines

1. Because you will use hydrochloric acid to test rocks, it is necessary that you wear safety goggles and protective gloves during this laboratory exercise any time you are using the acid solution.
2. Only use the hydrochloric acid if you think the mineral might be limestone or marble. One drop of hydrochloric acid solution is enough for our purposes.
3. Do not taste samples, even to test for halite.

Rock Identification

Your teacher may provide you with a laboratory data sheet for you to complete, or you may create your own table in your Lab Journal. The data sheets for igneous, sedimentary, and metamorphic rock identification are different. For each of the samples, it is usually best to work from left to right across the data sheet.

1. Igneous Rocks

   Use Tables 5.2 and 5.3 as aids to help identify minerals and name the igneous rock samples.

IGNEOUS ROCK IDENTIFICATION

Sample No.	Texture (coarse, fine, porphyritic, glassy, bubbles)	Color (light, intermediate, dark)	Minerals (list only minerals you can see)	Rock Name
1				
etc.				

Notes:

- For fine-grained rocks, the color is the most important property for rock identification. If you can see individual grains, then the minerals present are the most important property.

2. Sedimentary Rocks

   Use Table 5.5 as an aid to help identify the sedimentary rock samples.

SEDIMENTARY ROCK IDENTIFICATION

Sample No.	Grain Size and Description	Other Properties (acid test, crystals)	Rock Name
1			
etc.			

Notes:

- If individual mineral grains are not visible, the rock is siltstone, shale, non-fossiliferous limestone, halite, or coal.
- Sandstone fizzes lightly with a drop of hydrochloric acid if the sand grains are cemented together with calcite. Limestone fizzes much more vigorously than calcite-cemented sandstone.

3. Metamorphic Rocks

   Use Table 5.6 as an aid to help identify the metamorphic rock samples.

METAMORPHIC ROCK IDENTIFICATION

Sample No.	Foliated or non-foliated?	Description and Other Properties	Rock Name
1			
etc.			

Notes:

- Foliated rocks are at the top of Table 5.6. If the minerals are clearly lined up in the same direction or are in colored bands, the rock is foliated.

Chapter 6

Plate Tectonics and Mountain Building

Mount Everest, in the Himalaya Mountains on the border between China and Nepal, is Earth's highest mountain, with a summit 8,848 meters (29,029 feet) above sea level. Like most mountain ranges, the Himalayas are a complex mountain range, with a mixture of igneous, sedimentary, and metamorphic rocks spanning a range of ages. The summit of Mt. Everest has a layer of limestone containing an assortment of marine fossils. This limestone overlays older metamorphic rocks. Based on the fossils and the arrangement of the rock layers, geologists have concluded that the sedimentary rocks at the summit of Mount Everest were formed in a shallow sea that existed between India and the rest of Asia. Since that time, India has moved northward—look back at Figure 3.2—squeezing the sedimentary layers that had been deposited, resulting in the uplift of the marine rocks to their present elevation. India continues to move northward at about two centimeters per year, and the Himalayas are still uplifting at a rate of up to a centimeter per year. The motion of the continents is now described by the theory of plate tectonics, which has replaced the older theory of continental drift.

Objectives

After studying this chapter and completing the exercises, you should be able to do each of the following tasks, using supporting terms and principles as necessary.

1. Describe the evidence that first led people to consider that some of the continents might have once been joined together.
2. State the details of Wegener's four main arguments in favor of his theory of continental drift.
3. Explain why most geologists did not accept Wegener's theory of continental drift.
4. Summarize the evidence that led to the acceptance of seafloor spreading in the 1960s.
5. Explain how seafloor spreading and plate tectonics are different than Wegener's theory of continental drift.
6. Describe the three types of tectonic plate boundaries.
7. Explain how convection in Earth's mantle is related to plate tectonics.
8. Describe Earth's continental margins and seafloor.
9. Discuss the relationships of compression and tension to folding and faulting in Earth's crust.

Vocabulary Terms

You should be able to define or describe each of these terms in a complete sentence or paragraph.

1. abyssal plain
2. anticline
3. asthenosphere
4. atoll
5. continental drift
6. continental margin
7. continental rise
8. continental shelf
9. continental slope
10. convergent boundary
11. deep-ocean trench
12. divergent boundary
13. fault
14. fault-block mountain
15. fold
16. geomagnetic reversal
17. graben
18. guyot
19. horst
20. hydrothermal vent
21. manganese nodule
22. mid-ocean ridge
23. monocline
24. normal fault
25. Pangea
26. plate tectonics
27. reverse fault
28. seafloor spreading
29. seamount
30. stress
31. strike-slip fault
32. syncline
33. tectonic plate
34. thrust fault
35. transform boundary
36. volcanic island arc

6.1 Continental Drift

Within a century after the voyages of Christopher Columbus to the New World, the creation of reasonably accurate maps of the Atlantic Ocean led some scholars to note how the continents on opposite sides of the ocean, particularly South America and Africa, seemed to fit together like pieces of a puzzle, as illustrated in Figure 6.1. Abraham Ortelius, writing in the late 1500s, suggested that North and South America had torn away from Europe and Africa through a process of earthquakes and

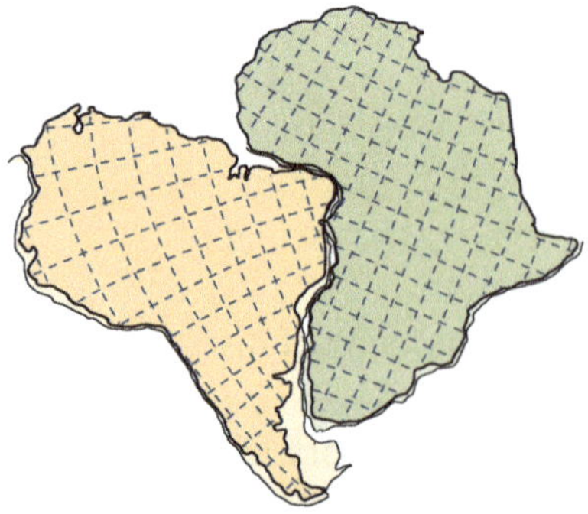

Figure 6.1. The fit between the coastlines of South America and Africa was noticed in the 1500s as soon as the first accurate maps of the Atlantic Ocean were created.

flooding. Others speculated about whether these continents had once been next to each other, but it wasn't until the early twentieth century that scientists began seriously to think about this hypothesis.

In 1915, a German meteorologist named Alfred Wegener published a book in which he presented multiple lines of evidence for a theory called *continental drift*. This theory proposed that not only the continents surrounding the Atlantic Ocean, but all the continents, were once joined together in a supercontinent called *Pangea*[1], illustrated in Figure 6.2. This supercontinent then split into a number of smaller continents, such as North America, Australia, and Antarctica. These continents then slowly moved to their present positions over a period of millions of years.

Wegener outlined four main arguments supporting the theory of continental drift:

1. *The fit of the continents.* Like others before him, Wegener was intrigued by the similarity of the coastlines on opposite sides of the Atlantic Ocean.
2. *Fossil evidence.* By Wegener's time, most paleontologists (scientists who study fossils and past life) believed that there must have been some sort of connection between Africa and South America in the past. In particular, fossils of an extinct aquatic reptile called *Mesosaurus* were found on both sides of the Atlantic, but nowhere else in the world. If *Mesosaurus* were a good enough swimmer to cross the Atlantic, then its fossils should be more widespread. Scientists had proposed that there had once been land connecting Africa and South America, but Wegener preferred to explain the distribution of *Mesosaurus* by continental drift. The distribution of a fossil fern called *Glossopteris*, which is found in all of the Southern Hemisphere continents, also suggested that these continents were once joined to each

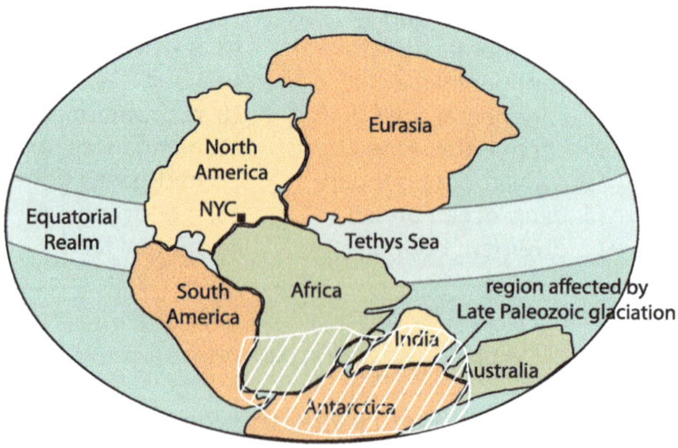

Figure 6.2. The arrangement of the continents about 250 million years ago.

1 Also called *Pangaea*. The name derives from the Greek *pan*, meaning "all" or "whole," and *Gaia*, meaning "Mother Earth" or "land."

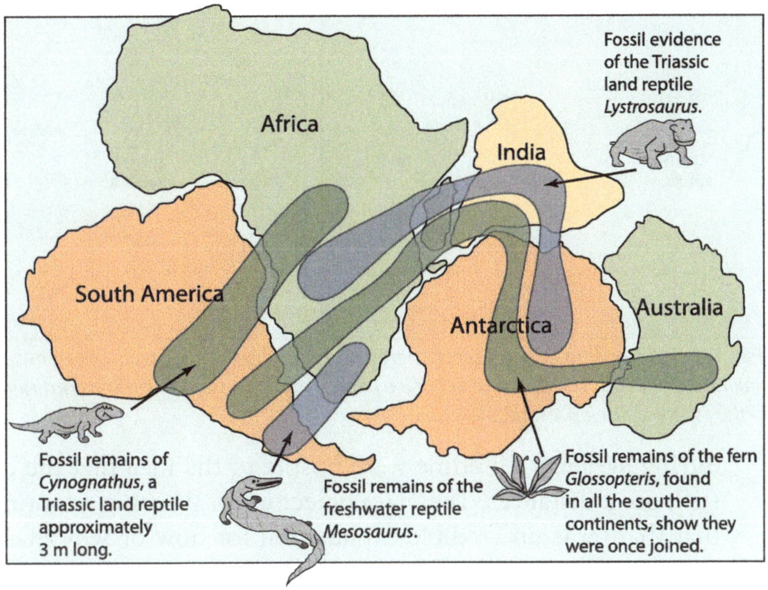

Figure 6.3. The distribution of fossils of the fern Glossopteris, *the aquatic reptile* Mesosaurus, *and other fossils, are evidence that the continents were once together in a supercontinent.*

other. The fossil evidence that the continents were once assembled together is summarized in Figure 6.3.

3. *Similarity of rock types.* On a jigsaw puzzle, not only do the shapes of the pieces have to fit together, the picture must be consistent from piece to piece as well. Wegener observed a similar matching of rock types across the Atlantic Ocean. The rocks of the Appalachian Mountains of eastern North America line up well with similar rocks in Great Britain and Scandinavia, as illustrated in Figure 6.4.

4. *Ice age deposits.* There is abundant evidence for an ancient ice age that covered parts of South America, Africa, India, Antarctica, and Australia. Geologists can look at scratches in the rocks and tell which way the glacial ice was moving, and the directions make most sense if these continents had once been connected to each other. Figure 6.5 illustrates how geologists believe these continents were connected during the ice age compared to their present positions.

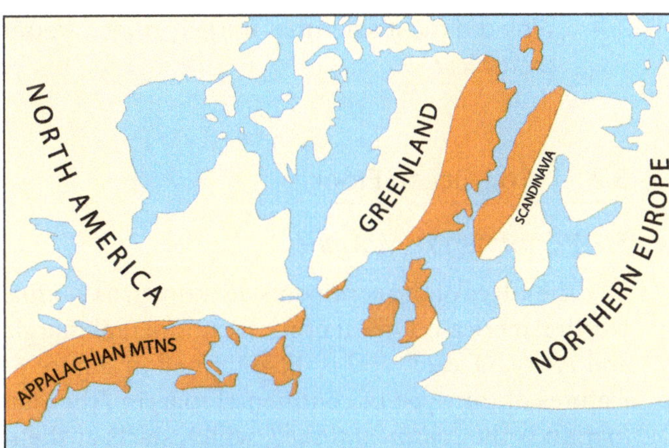

Figure 6.4. If the Atlantic Ocean did not exist between North America and Europe, rocks in mountain belts on the two continents would line up.

Despite multiple arguments in favor of continental drift, most geologists

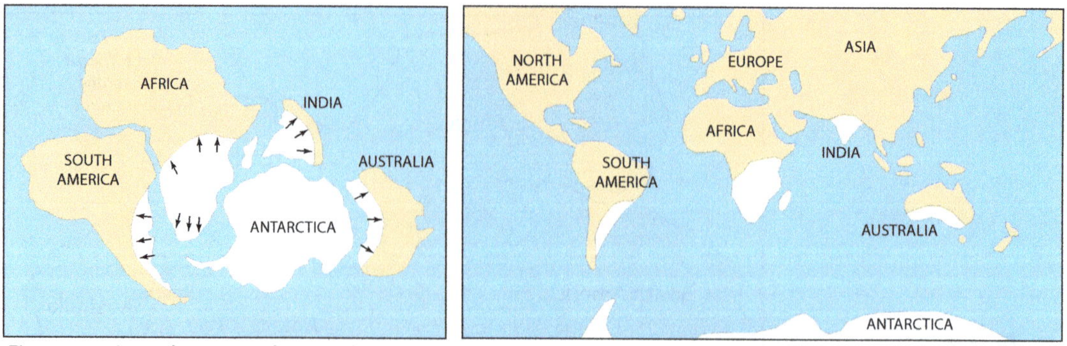

Figure 6.5. According to geologists, there was an ice age in the Southern Hemisphere between 220 and 300 million years ago. The distribution of ice makes much more sense if the continents were once all together than if they were in their present positions.

during Wegener's lifetime were hostile to the idea that the continents had moved such great distances. The main objection to Wegener's continental drift theory was that there was no credible explanation for how or why massive continents could move around on the face of the earth. It seemed that the continents would somehow have to plow through the rocky crust of the ocean floor in order to move around. Wegener perished in 1930 while on an expedition on the Greenland ice cap, and did not live to see his theory become widely accepted.

Learning Check 6.1

1. What was the earliest evidence that suggested that Earth's continents might have moved in the past?
2. How did glacial deposits in the southern continents such as Africa and Australia support the idea of continental drift?
3. How was fossil evidence used to support the theory of continental drift?
4. Why did most scientists initially reject Wegener's theory of plate tectonics?

6.2 The Ocean Floor

6.2.1 Continental Margins

The edges of the continents are known as *continental margins*. The east coast of North America is typical of most continental margins around the Atlantic, Indian, and Arctic Oceans, and is diagrammed in Figure 6.6. The figure shows the major features of this type of continental margin. The shallowest part of the continental margin is the *continental shelf*, which starts at the shoreline and extends outward anywhere from a few kilometers to several hundred kilometers. In most places, the continental shelf is a gently-sloping plain, and is underlain by continental crust, so continental shelves are really a part of the continents themselves, even though they

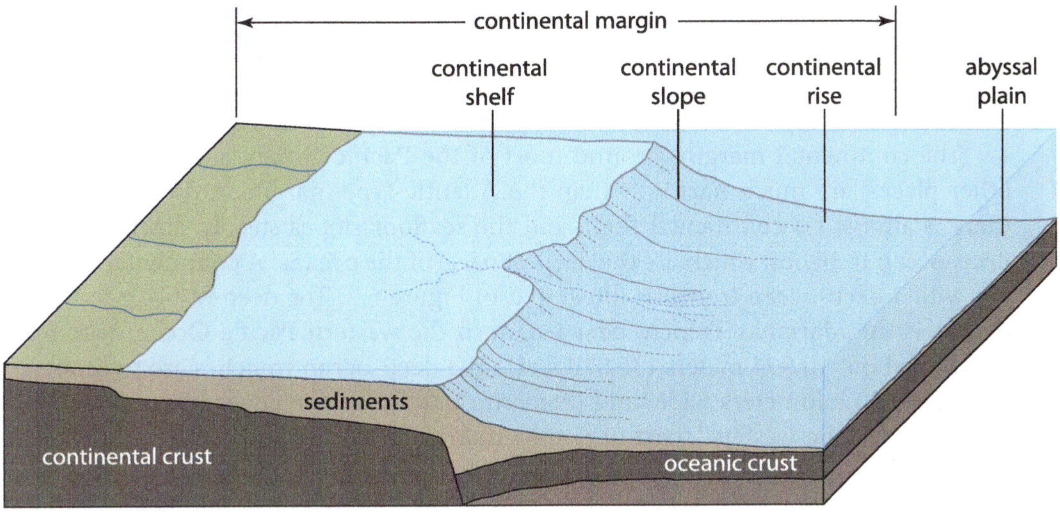

Figure 6.6. An Atlantic-type continental margin.

are under water. The continental shelf contains some of the world's greatest reserves of oil and gas, such as along the Gulf of Mexico coast of Texas and Louisiana. It is also the home of some important fisheries, such as the Grand Banks off the coast of Newfoundland, Canada.

The continental shelves extend out to a depth of around 130 meters. Beyond this, the seafloor drops more steeply down to the depths that are often 10,000 meters or greater. This sloping part of the seafloor is known as the *continental slope*. The continental slope is the true edge of the continents because it is where the granitic continental crust transitions to basaltic oceanic crust. In places, the continental slope is cut by deep canyons, which are carved by currents that carry sediments from the top of the slope down to deeper parts of the ocean. Along the base of the

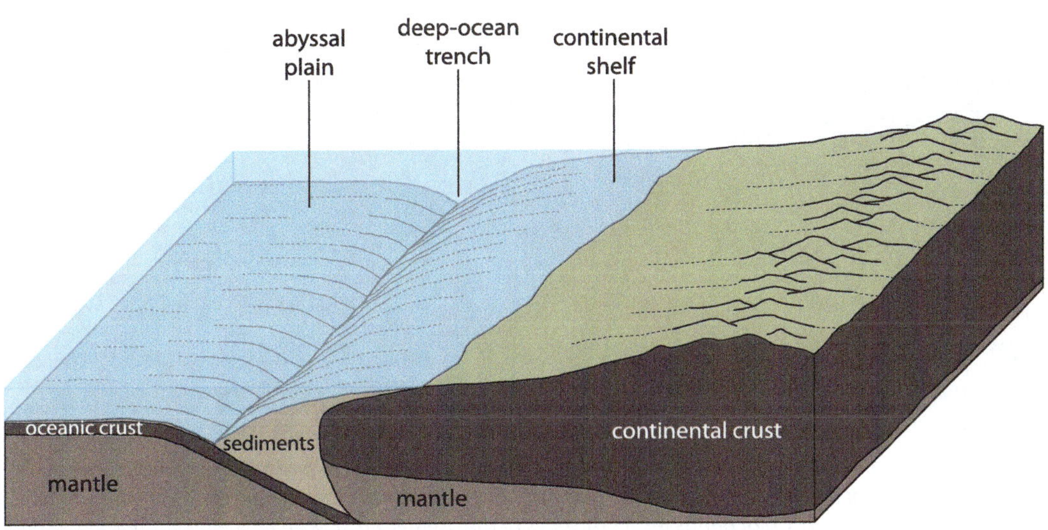

Figure 6.7. A Pacific-type continental margin with an offshore deep-ocean trench.

continental slope is a more gently-sloping region called the *continental rise*, which is made of sediments eroded and transported off of the continental shelf and slope. Beyond the continental slope lies the *abyssal plains*. These are blanketed by thick accumulations of fine mud and are among the flattest places on Earth.

The continental margins around most of the Pacific Ocean, as well as a few other places, are much narrower than the Atlantic-type margins. In some places there is almost no continental shelf, and the seafloor slopes steeply down to the deep-ocean trenches, which are the deepest parts of the oceans. A continental margin with a deep-ocean trench is illustrated in Figure 6.7. The deepest place in the oceans is the Marianas Trench, near Guam in the western Pacific Ocean, with a depth of about 10,994 meters (36,070 feet). The deep-ocean trenches are typically less than 100 kilometers wide, and often run parallel to the edges of continents, such as along the western coast of South America, or along arc-shaped chains of volcanic islands, such as along the Aleutian Islands of Alaska.

6.2.2 Mid-Ocean Ridges

A prominent feature of the ocean floor is the series of *mid-ocean ridges*, a mountain range that runs through all of the oceans, and has a combined length of about 70,000 kilometers. One portion of the mid-ocean ridge is the Mid-Atlantic Ridge, which runs down the center of the Atlantic Ocean, mirroring the coastlines on either side of the ocean, as shown in Figure 6.8. For the most part, the mountains

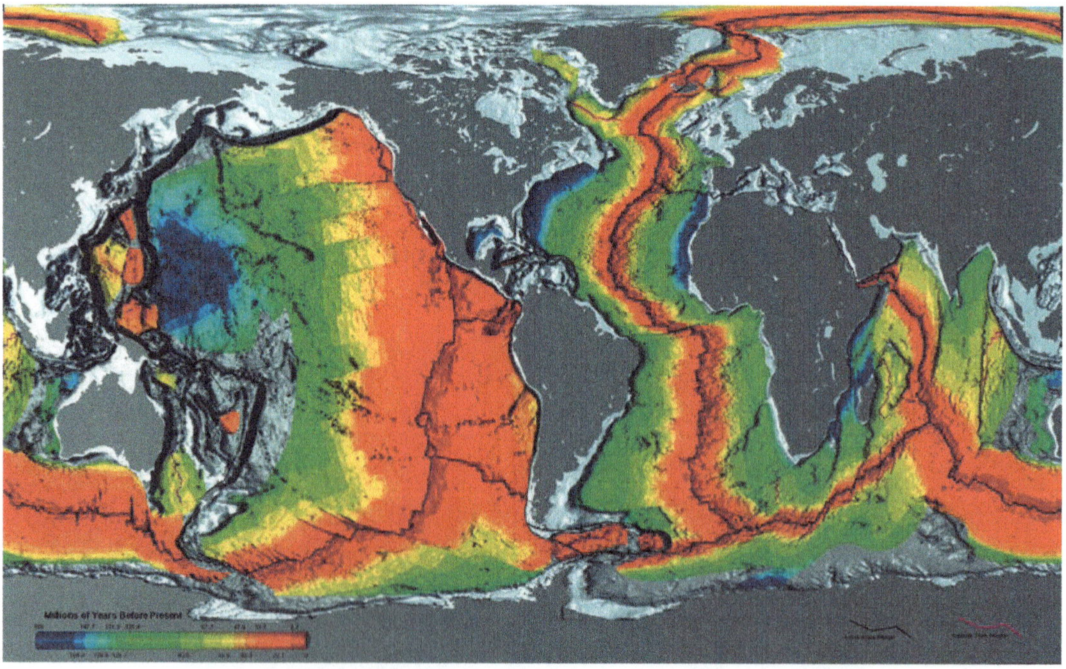

Figure 6.8. Earth's mid-ocean ridges. The ridges are the red and orange areas, and the black lines mark the rift zone along the axis of the ridges. Note how the Mid-Atlantic Ridge mirrors the coastlines on the east and west sides of the Atlantic Ocean. Color bands reflect the ages of the oceanic crust, with red being youngest and dark blue being oldest.

of the mid-ocean ridges fall well short of being tall enough to rise above sea level. One important exception is Iceland, which sits right on top of the mid-ocean ridge.

Figure 6.9. A "black smoker" is a type of hydrothermal vent on the ocean floor. The black cloud is composed of superheated water filled with tiny particles of sulfide minerals.

Along the center of much of the mid-ocean ridge system, there is a valley called a rift, which can be over 1,000 meters deep. This rift is the site of frequent volcanic activity. It is also the location of underwater hot springs called *hydrothermal vents*, which discharge superheated water with temperatures up to 350°C. One of these hydrothermal vents is shown in Figure 6.9. This fluid is rich in metal sulfides and oxides, which turn into crystals of minerals such as pyrite, chalcopyrite, and hematite. Underwater deposits formed by hydrothermal vents may someday be mined for the metals copper, zinc, lead, and others. Some ores now on the continents are believed to have formed in similar oceanic environments.

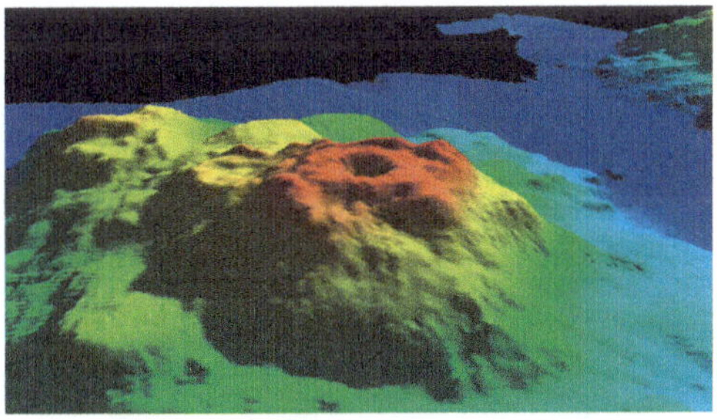

Figure 6.10. Ely Seamount in the Gulf of Alaska. Note the volcanic crater at the summit. This image and the one in Figure 6.11 were created from echo sounding data, as described in Figure 2.27.

6.2.3 Seamounts and Marine Sediments

Scattered about over much of the ocean floors are isolated volcanic mountains called *seamounts*. Figure 6.10 shows a seamount in the Gulf of Alaska. Seamounts number in the tens of thousands, and many rise hundreds or thousands of meters above the seafloor without breaking the surface of the ocean. Examples of seamounts that rise above sea level include the Hawaiian Islands and Tahiti in the Pacific Ocean, and the Azores and St. Helena in the Atlantic Ocean. Almost all seamounts have a volca-

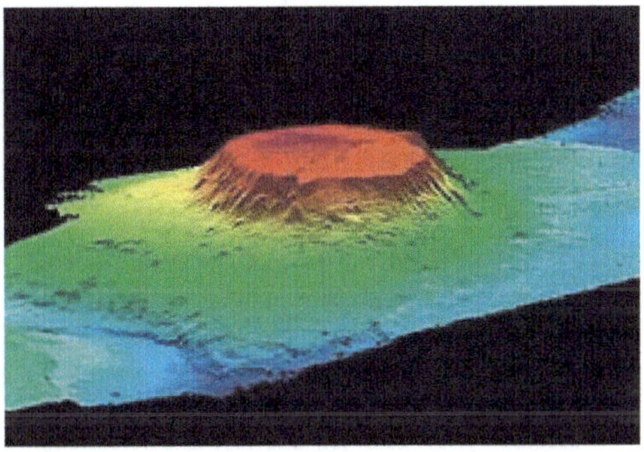

Figure 6.11. Bear Seamount, in the Atlantic Ocean off the coast of Massachusetts, an example of a guyot. The summit rises over 2000 meters above the seafloor.

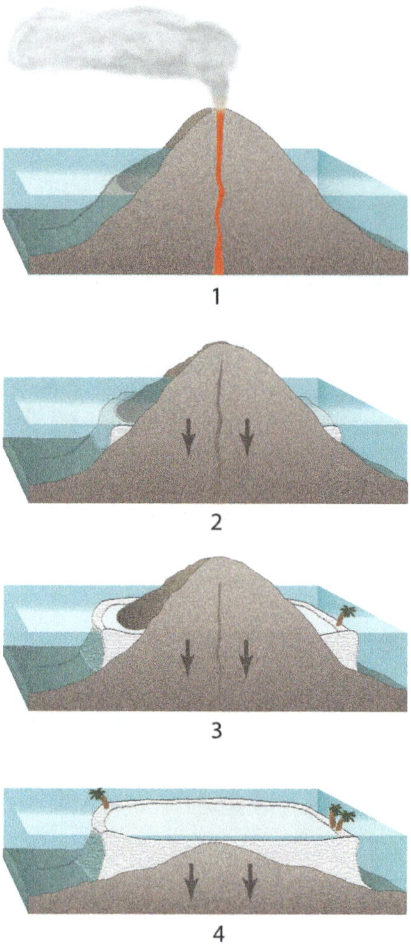

Figure 6.12. Progression from a volcanic island (1), to a fringing reef (2), barrier reef (3), to a coral atoll (4). If the island sinks further and coral growth stops, the next step would be a guyot.

Fringing reef, Maldives

Barrier reef around Bora Bora, Polynesia

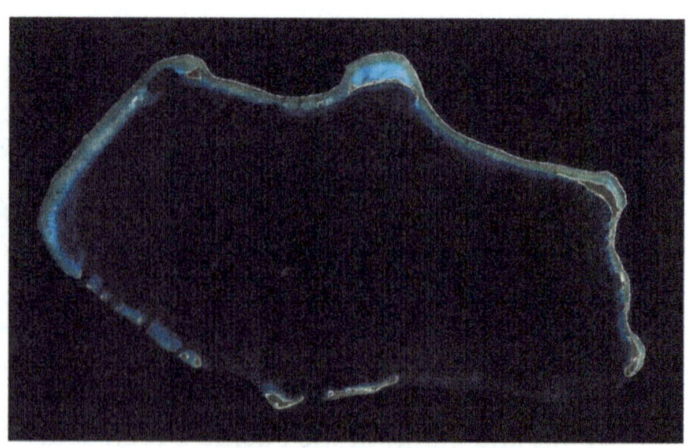

Bikini Atoll, South Pacific

Figure 6.13. Three islands illustrating progression of a sinking volcanic island through fringing reef, barrier reef, and atoll.

nic origin, and some are the sites of active volcanism.

A *guyot* is a flat-topped seamount, illustrated in Figure 6.11. The summits of guyots are hundreds or thousands of meters below the water surface, yet the sediments and fossils on their summits suggest that the summits were once at sea level. It is believed that these mountains once rose above

Figure 6.14. The White Cliffs of Dover in England are made of chalk. The chalk is composed of microscopic skeletons of single-celled marine organisms.

sea level, but were eroded to a flat surface by wave erosion. The remains of the volcano then subsided deeper into the ocean basin.

Guyots seem to be the end product of a process that happens to many volcanic islands. At some point, volcanoes in the oceans become extinct; that is, they stop erupting. In tropical waters, coral reefs will form around the islands. Coral organisms live as shallow-water colonies composed of many billions of individual organisms in tropical and sub-tropical waters, and produce skeletons made of calcium carbonate. When the reef is adjoining the island, it is called a fringing reef. Since the island no longer grows larger through volcanic eruptions, the island begins both to erode and sink. However, the coral is still able to grow, even as the rest of the island sinks. Eventually the reef makes a ring called a barrier reef, with the remnants of the volcano in the center. As the volcano continues to erode and subside, it gets to the point where the volcano is completely submerged beneath the waves, resulting in the formation of an *atoll*—a low-lying, ring-shaped island that surrounds a central lagoon. Figure 6.12 illustrates the progression from a volcano with a fringing reef to an atoll, and Figure 6.13 shows tropical islands at various stages in this process.

Sediment accumulations on the seafloor are thickest near the continents, and very thin at the mid-ocean ridges. There are three types of sediments found in deep ocean basins, and these formed in three different ways. The first is sediments that are derived from erosion of land. Sand and gravel tend to be deposited close to the shoreline, but extremely fine-grained clay can remain suspended in the water for many years. Scientists estimate that it can take between 5,000 and 50,000 years for just one centimeter of land-derived sediment to accumulate on the abyssal plains.

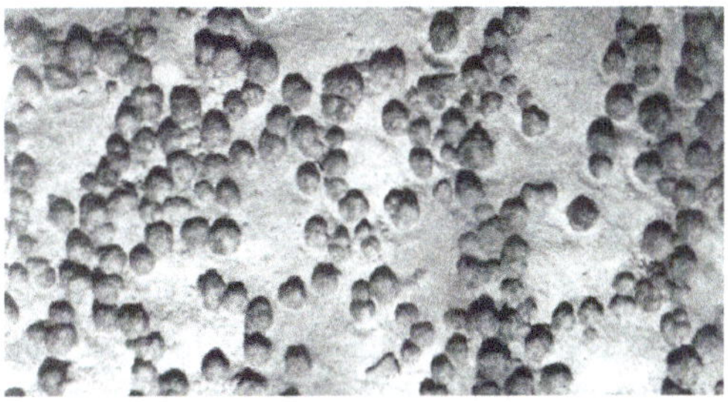

Figure 6.15. Manganese nodules on the seafloor.

Another source of sediments in the ocean basins is biological activity. Single-celled organisms in shallow seawater form microscopic skeletons made of either calcium carbonate or silica. When the organisms die, these skeletons settle to the seafloor, forming fine-grained sediments called calcareous ooze or siliceous ooze. Chalk, shown in Figure 6.14, is a type of limestone made of calcareous ooze. The chalk used on chalkboards used to be made of the rock chalk, but is now commonly made of the soft mineral gypsum.

Sediments formed directly from seawater represent a third type. An important example of this is *manganese nodules*, which are rocks that are about the size and shape of potatoes that lie scattered in many places on the ocean floor. Manganese nodules, shown in Figure 6.15, are made mostly of manganese and iron, but also contain significant amounts of valuable metals such as copper, nickel, chromium, molybdenum, and cobalt. Manganese nodules grow around some sort of nucleus, such as a sand grain or shark tooth, and appear to grow extremely slowly—at rates measured in millimeters per million years. The metals that go into these rocks come from ions dissolved in seawater. Manganese nodules have not been mined extensively to date, but if the price of the metals these nodules contain increases significantly, it might someday become economical to mine them.

Learning Check 6.2

1. Describe the different parts of the typical continental margin such as is found around most of the Atlantic Ocean.
2. How are the continental margins around much of the Pacific Ocean different than continental margins around the Atlantic Ocean?
3. Why are hydrothermal vents associated in many cases with the mid-ocean ridges?
4. Why do guyots generally have flat tops?
5. What are three different ways in which seafloor sediments form?

6.3 Seafloor Spreading

Another significant discovery about Earth's seafloor in the 1960s had to do with Earth's magnetic field. Because Earth's core is mostly iron, Earth behaves as a giant magnet. All magnets have a north pole and south pole, and Earth has a North Magnetic Pole and a South Magnetic Pole. (Earth's South Magnetic Pole is in the Northern Hemisphere, since the north pole of compasses point toward it.) An unexpected discovery made in the early 1960s was that Earth's magnetic field occasionally *changes polarity*, which means that the North Magnetic Pole becomes the South Magnetic Pole, and the South Magnetic Pole becomes the North Magnetic Pole! This change in the direction of Earth's magnetic field, illustrated in Figure 6.16, is called a *geomagnetic reversal*. The planet itself does not flip over, but something happens in the core that reverses the polarity of Earth's magnetic field. The

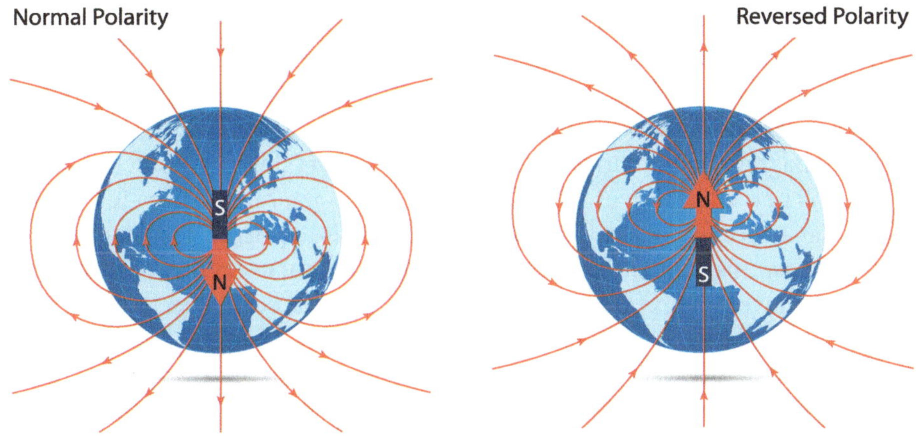

Figure 6.16. Earth's magnetic field reverses in polarity, typically every few hundred thousand years.

history of geomagnetic reversals is recorded in volcanic rocks. When magma crystallizes, iron-bearing minerals such as magnetite in the resulting rocks become oriented to Earth's magnetic field. In other words, these minerals act as compasses, pointing north and locking in place the orientation of Earth's magnetic field at the time the rock cooled. We now know that the last major reversal was about 700,000 years ago. Volcanic rocks formed since that time have "normal" polarity. From 2.4 million years ago to 700,000 years ago, most volcanic rocks have "reversed" magnetic polarity.

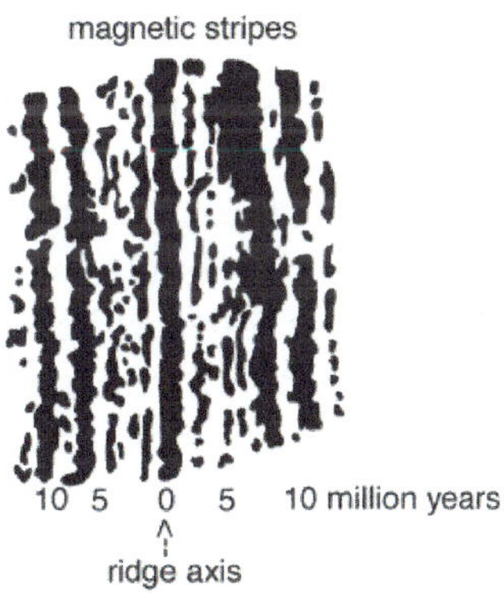

Figure 6.17. Bands of normal (black) and reversed polarity on the seafloor southwest of Iceland.

As oceanographers studied the seafloor in the 1960s, they discovered that the seafloor has bands of normal and reversed magnetic polarity preserved in the rocks, and that these bands run parallel to the mid-ocean ridges, as illustrated in Figure 6.17. The rocks right along the rift zone of the mid-ocean ridges have normal polarity. Rocks somewhat down the flanks of the ridges have reversed polarity, and rocks even farther from the ridge crest have normal polarity. Geologists interpreted this as meaning that the rocks along the ridge were the youngest, and that the seafloor gets progressively older with distance from the ridge.

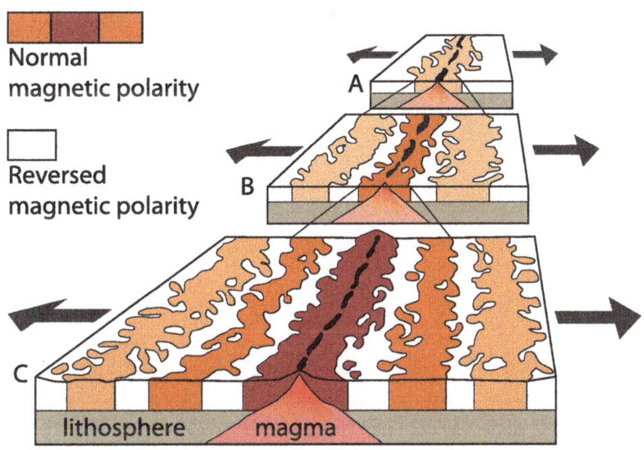

Figure 6.18. As magma forms new continental crust at the mid-ocean ridge, the newly-formed rocks have a magnetic polarity, either normal or reversed. Older crust moves away from the ridge.

In the early 1960s, geologist Harry Hess combined this knowledge of seafloor topography and seafloor magnetism into a theory called *seafloor spreading*. According to the theory of seafloor spreading, new oceanic crust forms at the mid-ocean ridges and then slowly moves away from the ridge crest. This new crust is formed by magma that intrudes up into fissures and crystallizes along the center of the ridge. As new crust moves away from the ridge—at a rate of just a few centimeters per year—room is made for more magma to rise up from the mantle. Figure 6.18 shows how this explains the magnetic banding of rocks on the seafloor. As new crust forms above a magma chamber beneath the mid-ocean ridge, it bears the current magnetic polarity, whether normal or reversed. As additional oceanic crust is formed at the mid-ocean ridge, older crust moves horizontally away from the ridge.

Seafloor spreading doesn't just involve Earth's crust, but Earth's mantle as well. For one thing, the upper mantle is the source for magma that gets pushed up into the mid-ocean ridges—a key element of seafloor spreading. Additionally, much of the mantle seems to be directly involved in the movements of Earth's crust. The crust plus the upper mantle, as you recall from Chapter 1, is referred to as the lithosphere; the deeper mantle is the *asthenosphere*. The lithosphere is fairly rigid, while the asthenosphere, though composed of solid rock, is able to flow because of the immense pressures at depth. You can think of oceanic lithosphere as resting upon a pair of conveyer belts, which move oceanic lithosphere away from the mid-ocean ridge. This process is illustrated in Figure 6.19. As magma rises near the ridge, older oceanic lithosphere is carried away from the ridge. As new oceanic crust is formed at the mid-ocean ridges, old oceanic crust is recycled back into the mantle at the deep-ocean trenches.

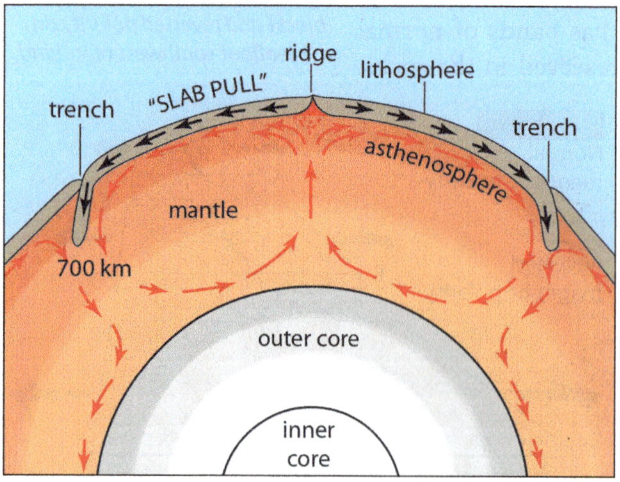

Figure 6.19. According to the theory of mantle convection, seafloor spreading occurs because of convection currents in the asthenosphere, which is the deeper part of Earth's mantle. This convection involves rising hot mantle material beneath mid-ocean ridges, and sinking mantle material near the deep-ocean trenches.

Seafloor spreading occurs at a rate of only a few centimeters per year, but it has been measured. Scientists use lasers to measure distances between continents. The light in a laser beam travels at the speed of light, about 300,000 km per second. By timing how long it takes laser beams to bounce off satellites in Earth orbit—measurements with precision in trillionths of seconds—scientists are able to confirm the motions of the continents. The motions of the continents have also been confirmed by measurements made using sophisticated GPS receivers. There are additional evidences for seafloor spreading, some of which we will look at in the next chapter on volcanoes and earthquakes.

Learning Check 6.3

1. Describe features of the mid-ocean ridges that support the theory of seafloor spreading.
2. How do patterns of geomagnetism recorded in rocks of the seafloor support the theory of seafloor spreading?
3. In what ways is the mantle believed to be involved in seafloor spreading?

6.4 Plate Tectonics

Seafloor spreading provides the needed mechanism to explain Alfred Wegener's hypothesis of continental drift. Geologists no longer use the term continental drift to describe the motions in Earth's crust, because it is not just the continents that are in motion. Instead, geologists now speak of the theory of *plate tectonics*[2]. According to the theory, Earth's entire lithosphere—including the crust and uppermost part of the mantle—is composed of *tectonic plates* that are slowly moving around on Earth's surface. Plate tectonics is one of the most important concepts in

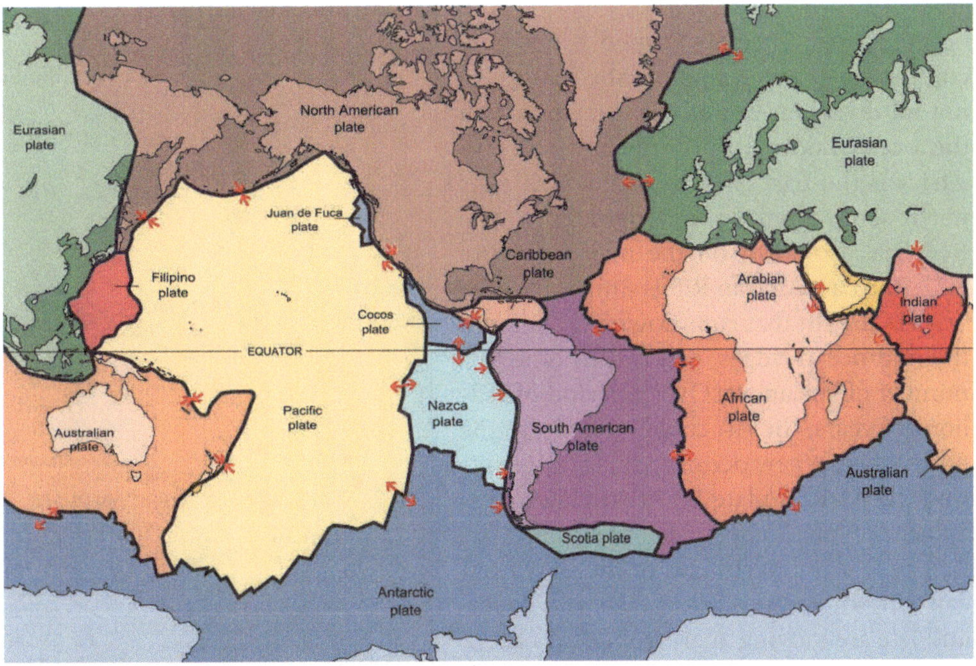

Figure 6.20. Earth's lithosphere is composed of seven major plates and a number of smaller plates.

2 "Tectonics" refers to processes that deform Earth's crust.

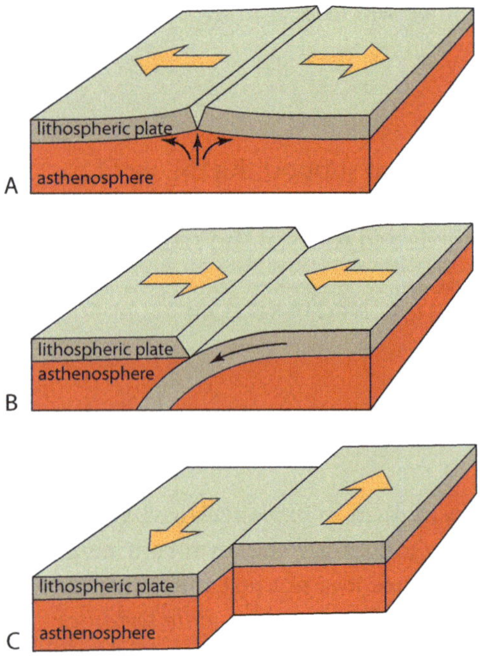

Figure 6.21. The three main types of tectonic plate boundaries: A. divergent, B. convergent, C. transform.

geology. The theory explains a wide variety of phenomena, such as most of the world's volcanoes and earthquakes, and the formation of mountain ranges and many ore deposits.

According to the theory of plate tectonics, Earth's lithosphere is divided into seven major plates, and a number of smaller plates. As you can see in Figure 6.20, the North American Plate includes not only most of the continent of North America, but also a significant amount of the Atlantic and Arctic Oceans. The North American Plate, like the other plates, moves around on Earth's surface as a solid unit.

6.4.1 Plate Boundaries

There are three types of tectonic plate boundaries, shown in Figure 6.21. The first of these, called *divergent boundaries*, are where plates are moving apart from each other. Most divergent boundaries are found at the mid-ocean ridges, where new oceanic crust is being formed by magma coming up from Earth's mantle. However, not all divergent boundaries are found on the ocean floors. A prominent feature in Africa is the East African Rift, which runs north-south from Lake Malawi up to the Red Sea, as shown on the map in Figure 6.22. The East African Rift is similar in some ways to the rift that runs down the center of the mid-ocean ridges, and is the home to a number of volcanoes. Over a period of millions of years, the rift could open up to the point where a new ocean basin forms, splitting the African plate into two plates. This appears to have happened already along the Red Sea, at the north end of the East African Rift, where the African Plate and Arabian Plate are moving apart from each other.

A second type of plate boundary is the *convergent boundary*, where two plates col-

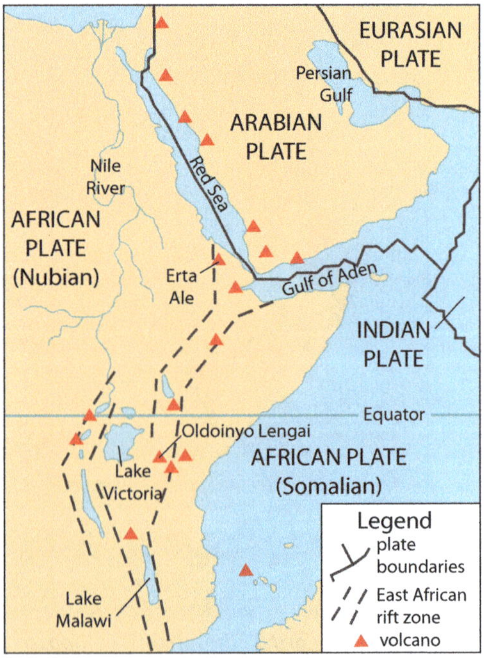

Figure 6.22. The East African Rift, a divergent plate boundary on a continent.

lide with each other. There are three types of convergent boundaries, based on the types of crust involved.

1. *Oceanic-continental convergence.* When oceanic lithosphere collides with continental lithosphere, the oceanic lithosphere is forced downward into the mantle, usually at an angle of about 45°. The basaltic oceanic crust is forced downward because its density (3.3 g/cm³) is higher than the granitic continental crust density (2.7 g/cm³). This process, in which an oceanic plate dives down into the mantle beneath another plate, is called *subduction*. Subduction creates deep-ocean trenches, as illustrated in Figure 6.23, and is a process that recycles oceanic crust that was created at mid-ocean ridges. As oceanic lithosphere subducts deeper into Earth's mantle, it heats and eventually partially melts, providing magma for overlying volcanoes. One example of continental-oceanic convergence is along the west coast of South America, where the Nazca Plate is subducting beneath the South American Plate. This forms the Peru-Chile Trench, which runs along most of the western coast of South America.

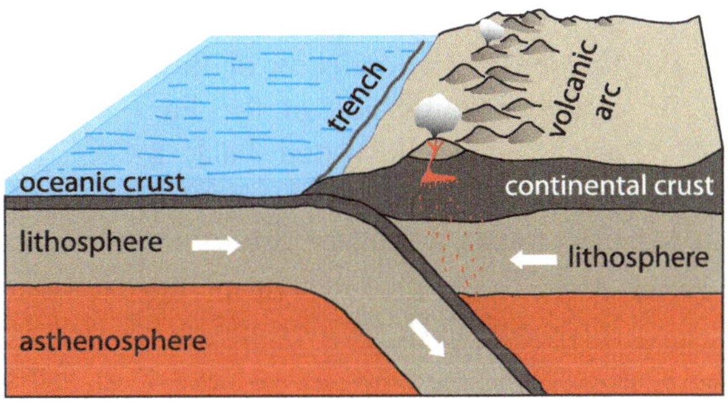

Figure 6.23. When an oceanic plate converges with a continental plate, oceanic lithosphere is subducted into Earth's mantle, and a deep-ocean trench is formed.

2. *Oceanic-oceanic convergence.* When oceanic crust converges with oceanic crust, one of the plates will subduct beneath the other, as illustrated in Figure 6.24. Typically, one of the plates is older and cooler than the other, and this is the plate that subducts. The subduction results in a deep-ocean trench in the middle of an ocean basin. An example of oceanic-oceanic convergence is at the Aleutian Trench in the North Pacific Ocean, where the Pacific Plate subducts beneath an oceanic portion of the North American Plate. The deep-ocean trenches associated with oceanic-ocean convergent boundaries are usually curved, and have a chain of volcanic islands parallel to the trench on the plate that is not subducted.

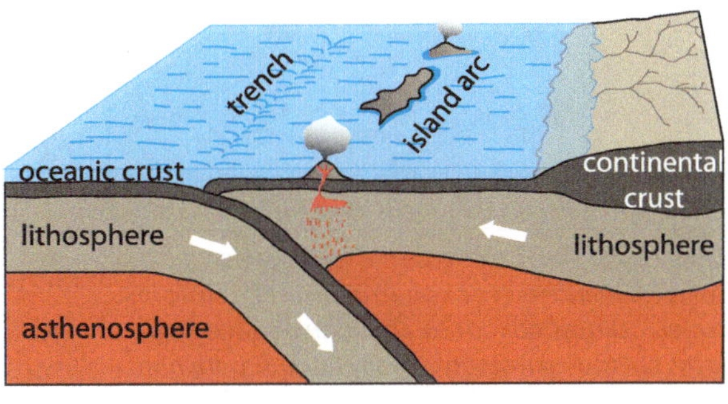

Figure 6.24. When an oceanic plate converges with another oceanic plate, one plate subducts beneath the other, and a deep-ocean trench forms. A line of islands called a volcanic island arc forms parallel to the trench.

3. *Continental-continental convergence.* When two continental plates collide, neither can be

subducted into the mantle because the low density of granite makes continental crust buoyant. As illustrated in Figure 6.25, when continents collide the crust is buckled and deformed, forming a mountain range. The Himalaya Mountains, the highest mountain range on Earth, formed by the collision of India into the south side of the Eurasian Plate, illustrated back in Figure 3.2. This process continues today. Another mountain range that formed by continental-continental convergence millions of years ago is the Appalachian Mountains, which formed when Africa collided with North America over a period of time between 325 and 260 million years ago. This collision was part of the events that joined the continents together to form the supercontinent Pangea, shown in Figure 6.2. Africa and North America later split, forming the Atlantic Ocean.

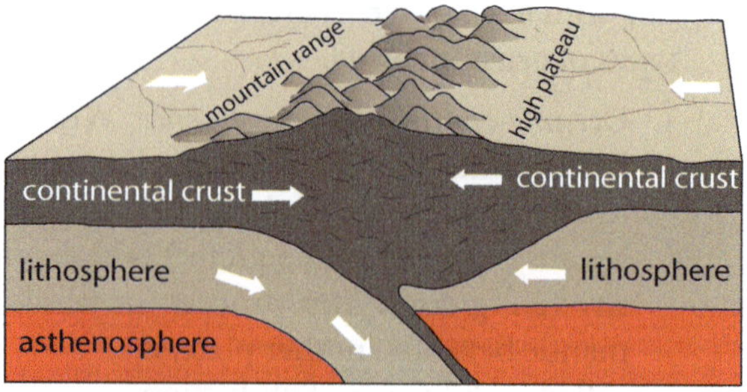

Figure 6.25. When continental plates converge, neither plate can be subducted into Earth's mantle. The Himalaya and Appalachian Mountains formed in this fashion.

The third type of plate boundary is the *transform boundary*, which involves two plates sliding past each other. At divergent boundaries, new crust is created; at convergent boundaries, old oceanic crust is destroyed, but at transform boundaries crust is neither created nor destroyed. Transform boundaries usually occur in oceanic crust, and are common along offset portions of the mid-ocean ridges. Figure 6.26 locates some transform boundaries along the western coast of North America, where the boundary between the Pacific Plate and the smaller Juan de Fuca Plate is made up partly of mid-ocean ridges (red lines) and partly of transform faults.

Figure 6.26. The Pacific coast of western North America has divergent, convergent, and transform tectonic plate boundaries. The Juan de Fuca Ridge is part of the mid-ocean ridge system. Subduction of the Juan de Fuca Plate beneath the North American Plate occurs without a visible deep-ocean trench off the coast of Washington and Oregon. The San Andreas Fault and the far eastern part of the Mendocino Fracture Zone are transform plate boundaries, where plates slide past one another.

There are only a few places where transform boundaries occur in continental crust. One of these is in California, where the Pacific Plate slides past the

North American Plate along the San Andreas Fault. The piece of California west of the San Andreas Fault is on the Pacific Plate, and is moving northward relative to the North American Plate. The Pacific coast of North America actually has convergent, divergent, and transform plate boundaries close to one another, as Figure 6.26 indicates.

6.4.2. Tectonic Plate Movements in the Past

If we project plate motion millions of years into the past, we can see that Earth was once a very different place. At present, the Atlantic Ocean is becoming 2 cm wider each year. If we were able to reverse this motion to some point in the past, the Atlantic Ocean would have been completely closed, with North and South America united with Europe and Africa. Doing the same thing with the motion of other plates, such as the Indian, Australian, and Antarctic Plates, leads us back to Wegener's idea of Pangea, the supercontinent that included most of Earth's landmasses over 100 million years ago.

We can do the same thing by looking into the future. What would a map of Earth look like millions of years from now? Would the Atlantic Ocean continue to open wider, with the Pacific Ocean getting narrower? Geologists believe there is evidence that the widening of the Atlantic Ocean will eventually slow down and even reverse, leading to an eventual re-closing of the Atlantic.

Learning Check 6.4

1. Describe the three types of plate boundaries.
2. What are the similarities and differences between the three types of convergent plate boundaries?
3. Describe what would happen to Africa in a few million years if divergence were to continue at the Great Rift Valley (the East African rift).
4. If the Atlantic Ocean is becoming 2 cm wider each year, how much wider was it in the year 2015 than it was in 1492, the year Columbus made his first trip across the Atlantic?

6.5 Mountain Building

Mountains can form in a number of different ways. For example, volcanoes, such as Mt. Rainier in Washington or Vesuvius in Italy, form when lava flows erupt from a central vent, growing to considerable height after dozens or even hundreds of eruptions. We will take a closer look at volcanoes and how they relate to plate tectonics in Chapter 7. Most mountain ranges are formed by the forces involved in plate tectonics, which can contort rock layers by squeezing them from the sides, or stretch rocks until they break apart. We will start by looking at the various ways mountain building processes can deform rocks.

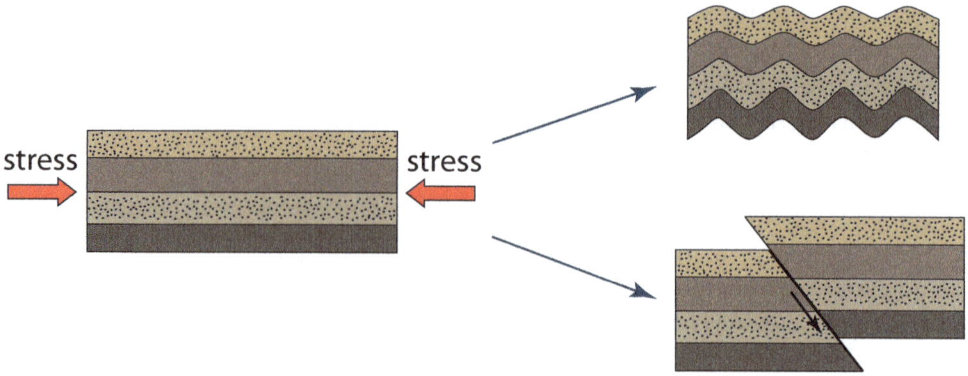

Figure 6.27. Compressional stress causes rocks to deform by folding or breaking (faulting), depending on the rocks and the conditions.

6.5.1 Folding

When tectonic plates converge with one another, a tremendous amount of force is applied on solid rocks in Earth's crust. This force is measured as *stress*, which is force per unit area. A low amount of stress can actually cause rock layers to bend rather than break, especially at the temperatures and pressures encountered deeper in Earth's crust. If the stress is applied for long periods of time, rock layers can become highly distorted. If rocks are subjected to high amounts of stress, it is more likely that the rocks will break rather than bend.

When rocks are squeezed by stress coming from opposite directions, the stress is called *compression*. Compression of rocks can result in either folding or breaking, as illustrated in Figure 6.27. A *fold* is a bend in a rock layer. Most sedimentary rocks are deposited in horizontal layers; if the layers are no longer horizontal, it can be assumed that the layers have been folded. There are three basic types of folds: *anticlines*, *synclines*, and *monoclines*, illustrated in Figure 6.28. An anticline, like the one shown in Figure 6.29, is a fold where rock layers slope away from the central axis of the fold, usually looking somewhat like an upside down "U". In some areas, the folded layers of sedimentary rocks in anticlines act as good petroleum traps, such as the top half of Figure 5.22. A syncline is a fold in which the rock layers slope inward toward the central fold axis, a right-side-up "U". Anticlines and synclines are usually found next to each other or in alternating patterns

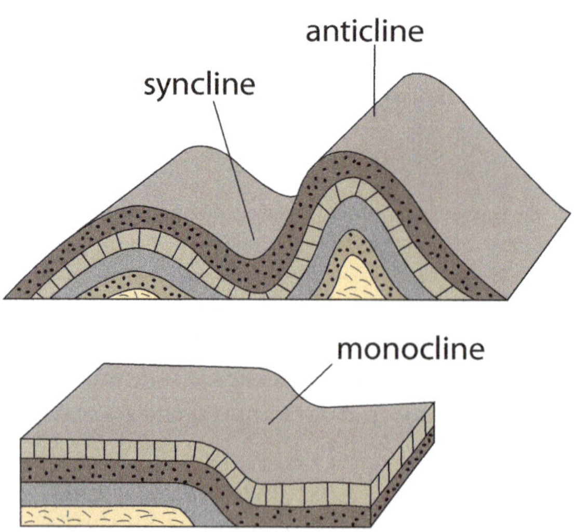

Figure 6.28. The three types of folds are anticlines, synclines, and monoclines.

of anticlines and synclines. The third type of fold is a monocline, commonly found in areas with mostly horizontal layers of sedimentary rocks. The monocline is like a step with horizontal rocks on either side of the fold.

Figure 6.29. An anticline in Lebanon.

Folds can be large or small features. The anticline in Figure 6.29 is large enough to form a mountain. On the other hand, the folds in Figure 6.30 are visible in a small outcrop of rock. Some folds are even smaller and are best studied in hand-sized samples or even with microscopes.

It might seem strange to you to think of solid rocks bending. Part of this is because the small samples we can hold in our hands seem rather inflexible. Another reason is because we are accustomed to observing rocks in the low-stress environment of Earth's surface. At the high pressures and temperatures deep in Earth's crust, rocks are much more likely to deform by folding than they are near the surface. But even near Earth's surface, solid rocks can fold. Geologists studying thin sections of folded rocks with microscopes can see where sand grains have moved past one another as the rocks folded, or tiny fractures that cut through the rock, enabling the rock to fold. Many folded sedimentary rocks also show evidence of fracturing that occurred along

Figure 6.30. These anticlines and synclines in southern Greece are each just a few meters across. The size of folds ranges from many kilometers down to millimeters.

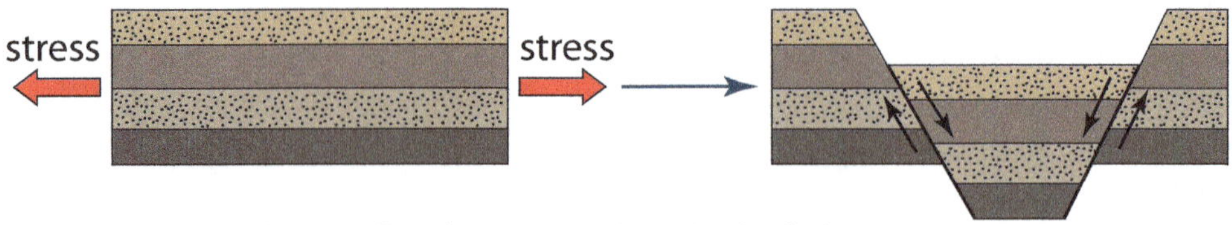

Figure 6.31. Tensional stress pulls rocks apart, which can lead to faulting.

with the folding. Folding can occur in sedimentary, metamorphic, and igneous rocks.

6.5.2 Faulting

Compressional stress can cause rocks to fold, as we have seen, but can also cause rocks to break, as we saw in Figure 6.27. A *fault* is a break in rocks along which movement has occurred. If the rock cracks, but there is no movement along the crack, the crack is considered to be a *fracture*, but not a fault. The opposite of compression is tension, which is stress that stretches rocks apart, rather than squeezing them together, as illustrated in Figure 6.31. Tensional stress occurs at divergent plate boundaries, such as at the East African Rift mentioned earlier, and also at other places within tectonic plates. Tension can also cause rocks to break, creating faults.

If a fault is not vertical in Earth's crust, geologists refer to the rocks above the fault as the *hanging wall*, and the rocks below the fault as the *footwall*, as illustrated in Figure 6.32. If you can imagine standing inside Earth's crust with your feet right on the fault, your feet would be on the footwall, and the rocks hanging above your feet would be the hanging wall.

If the hanging wall moves down relative to the footwall, the fault is called a *normal fault*. Normal faults are caused by tension in Earth's crust; the faults in Figure 6.31 are normal faults. If the hanging wall moves up relative to the footwall, the fault is known as a *reverse fault*, which is the type of fault that was shown in Figure 6.27. Reverse faults are caused by compression in Earth's crust.

Some reverse faults have a low angle relative to a horizontal plane, and are known as *thrust faults*. Like all reverse faults, thrust faults form because of compressional stress, such as what is found at convergent plate boundaries. Thrust faults can even be horizontal, sliding older rocks over younger rocks. One of the most famous examples of this is found in Glacier National Park in Montana, where older rocks have been pushed about 80 km (50 miles) horizontally over younger rocks, as illustrated in Figures 6.33 and 6.34. Normally, the oldest rocks

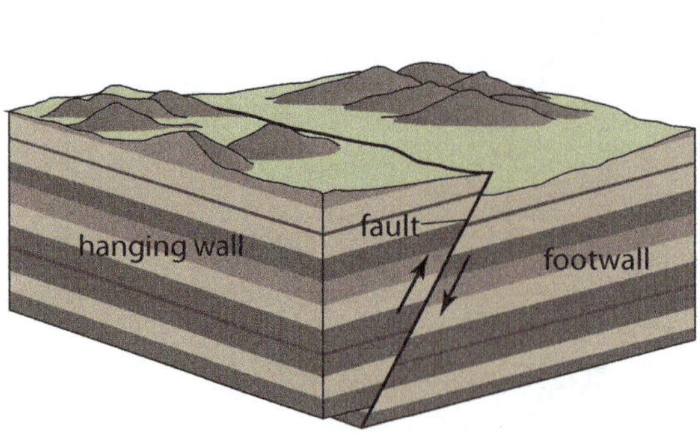

Figure 6.32. The hanging wall and footwall of a fault. To an ant on the fault, the foot wall would be beneath their feet, and the hanging wall would be hanging above them.

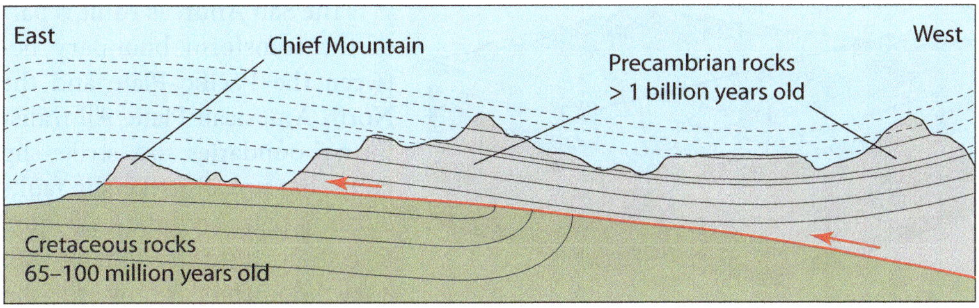

Figure 6.33. Thrust faulting occurs when compression causes older rocks to slide over younger rocks. In this diagram of the main thrust fault in Glacier National Park, Montana, Precambrian sedimentary rocks that are over one billion years old have been thrust over Cretaceous sedimentary rocks, which are about 80 to 110 million years old. Solid lines show layering in the rocks; dashed lines indicate probable layering before erosion in the mountains.

are deepest in Earth's crust, and the youngest rocks are nearer the surface. A thrust fault can stack rocks so they are "out of order," with the oldest rocks above younger rocks.

Before the development of the theory of plate tectonics, geologists had no explanation for how such dramatic compressional forces could exist in Earth's crust. The convergence of tectonic plates provides a source for these powerful compressional forces. Thrust faults are forming today in places like the Andes Mountains in South America, where oceanic-continental convergence is occurring, and in the Himalaya Range in Asia, which is the site of ongoing continental-continental convergence.

Another important type of fault is the *strike-slip fault*, where movement along the fault is horizontal rather than vertical, as illustrated in Figure 6.35. The San Andreas Fault, source of many of California's earthquakes, including the 1906 earthquake that leveled San Francisco, is a strike-slip fault. The San Andreas Fault is mapped on Figure 6.26.

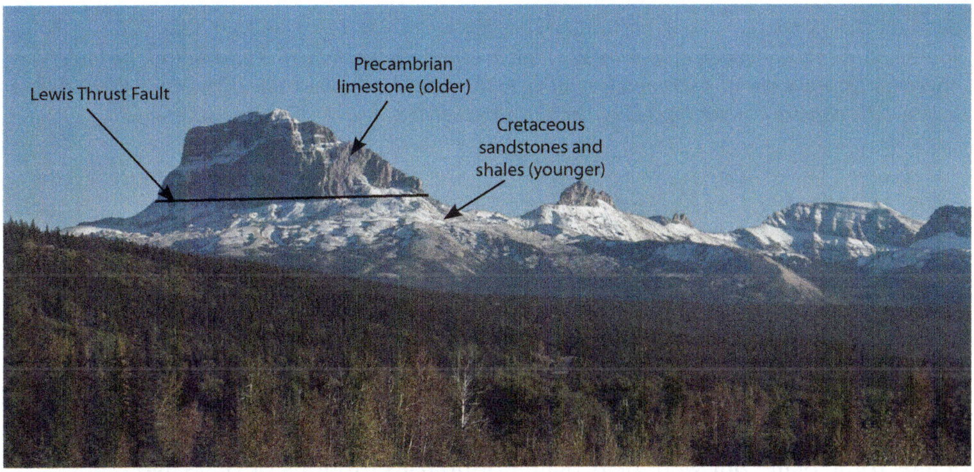

Figure 6.34. Chief Mountain in Glacier National Park, with older (Precambrian) rocks pushed over younger (Cretaceous) rocks along a nearly horizontal thrust fault.

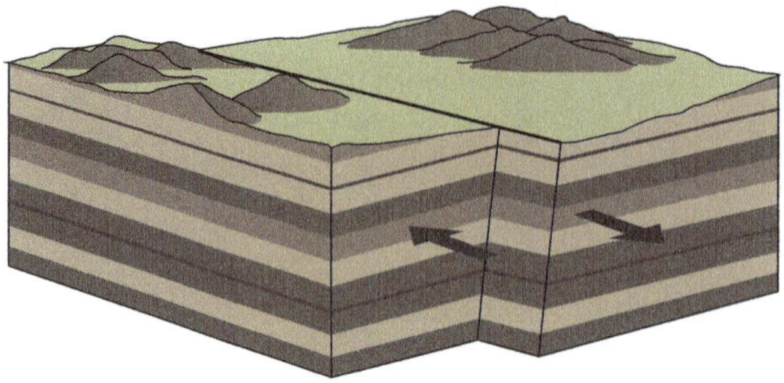
Figure 6.35. Movement along a strike-slip fault is horizontal.

The San Andreas Fault is part of the transform boundary between the Pacific Plate and the North American Plate. All transform boundaries are strike-slip faults, but not all strike-slip faults occur at plate boundaries. Figure 6.36 shows a strike-slip fault with a few kilometers of movement in the middle of the Eurasian plate.

Figure 6.36. A satellite image of a strike-slip fault in China. The rocks on the east side of the fault have moved horizontally about 4 kilometers relative to the rocks on the west side of the fault. (Note that in this view the camera is pointing straight down, although it may at first appear otherwise.)

6.5.3. Mountains Formed by Folding, Faulting, and Regional Uplift

Most mountain ranges on Earth can be related one way or another to plate tectonics. The movement of tectonic plates provides the energy necessary to cause uplift and deformation of vast areas of Earth's crust. The Rocky Mountains are near the western edge of the North American Plate, and the Alps are related to the boundary between the Eurasian Plate and the African Plate. Some mountain ranges, such as the Appalachian Mountains of the eastern United States, are not now near the edge of a tectonic plate, but were near plate boundaries millions of years ago.

All mountain ranges were formed by a combination of processes, such as volcanism, folding and faulting, but some mountain ranges are dominated by one of these processes over the others. The Appalachian Mountains are a mountain range where folding was an especially important part of their formation. Figure 6.37 shows a satellite image of part of the Appalachians. In Figure 6.38, a syncline is exposed in a road cut along a highway in the Appalachian Mountains.

Figure 6.37. A satellite image of a portion of the Appalachian Mountains in Pennsylvania. The darker colors are forested areas on ridges formed by anticlines and synclines.

Figure 6.38. This road cut through one of the ridges of the Appalachian Mountains in Maryland provides an excellent view of a syncline.

Fault-block mountains form when faulting causes large blocks of Earth's crust to either rise or sink relative to one another, usually along normal faults. In fault-block mountains, uplifted blocks are called *horsts*, and blocks that have been downthrown are called *grabens*, as you can see in Figure 6.39. Many of the mountain ranges of the western United States, such as the Sierra Nevada in California, the Wasatch Range of Utah, and the Basin and Range area of Nevada and surrounding states, are fault-block mountains. Figure 6.40 shows the Grand Tetons of Wyoming. There has been up

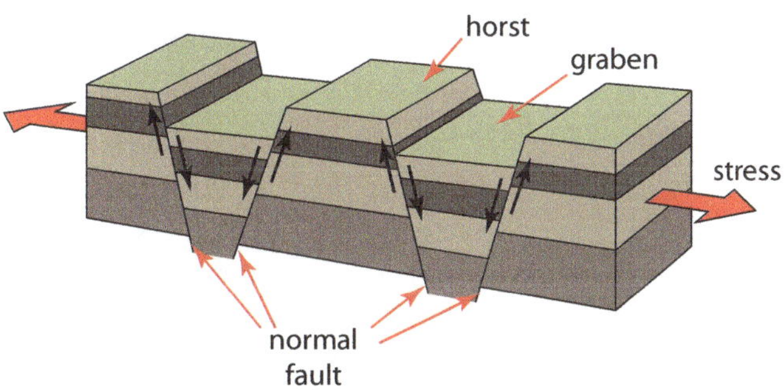

Figure 6.39. Fault-block mountains with uplifted blocks called horsts, and downthrown blocks called grabens.

Figure 6.40. The Grand Tetons of Wyoming. The mountain range is a horst, and the valley in the foreground is a graben.

to 9,000 meters (30,000 feet) of upward movement along the Teton Fault at the base of the Grand Tetons. Most of the uplifted block has been eroded away, and the graben has been filled with thousands of meters of sediments, leaving the tops of the mountains about 2200 meters above the valley floor.

Other mountain ranges are formed by regional uplift, in which broad areas of Earth's crust are lifted to higher elevations. The Colorado Plateau of the southwestern United States is an example of an area that has been uplifted in this manner. Much of the Colorado Plateau has been uplifted by over a kilometer with relatively minor folding and faulting of sedimentary rock layers, as evidenced by the nearly horizontal layers seen in places like the Grand Canyon, shown in Figure 3.8.

Learning Check 6.5

1. Describe different ways in which stress can deform rocks.
2. Explain why compressional stress can cause rocks either to fold or fault, while tensional stress generally causes only faulting.
3. Draw sketches of an anticline, syncline, and monocline.
4. Draw sketches of a normal fault, reverse fault, and strike-slip fault.
5. How are folding and faulting related to plate tectonics?

Chapter 6 Exercises

Answer each of the questions below as completely as you can. Write your responses in complete sentences unless instructed otherwise.

1. The far eastern coast of Canada and the west coast of Ireland are about 3000 kilometers apart. If the Atlantic Ocean is widening by about 2 centimeters each year, how many years did it take for the Atlantic Ocean to open? Assume a constant rate of opening (which may or may not be a reasonable assumption).
2. Which of Wegener's arguments for continental drift seems strongest to you?
3. Draw and label a series of diagrams illustrating the development of an atoll.
4. The magnetic bands on the seafloor near the East Pacific Rise, part of the mid-ocean ridge system, are farther apart than the magnetic bands associated with the Mid-Atlantic Ridge. Formulate a hypothesis to explain the cause of this.
5. Explain why seafloor sediments are thinner near the mid-ocean ridge than they are in the abyssal plains.
6. Explain how mantle convection is related to mid-ocean ridges and deep-sea trenches.
7. Thinking again about our discussion of theories in Chapter 3, how would you respond to someone who says, "I don't believe in plate tectonics because plate tectonics is just a theory?"
8. What types of plate boundaries are most often associated with mountain ranges on continents? Why?
9. What types of faulting would you expect to find in the Himalaya Mountains?
10. Geologists sometimes speak of projecting current processes like the motions of tectonic plates millions of years into the future. In light of the Christian hope of the return of Christ, and the creation of a New Earth written about in Revelation 21-22, how should Christians think about the future of Earth?
11. The portion of California west of the San Andreas Fault is on the Pacific Plate, and is moving north relative to the North American Plate at an average rate of 3.5 cm per year. At this rate, how many years would it take for western California to move up to the Gulf of Alaska, 2500 km to the north?

Chapter 7

Volcanoes and Earthquakes

In August of AD 79, a series of earthquakes shook the area around Vesuvius, a volcano in Italy and the largest volcano on the European mainland. The people of central Italy were used to earthquakes, so it seems most people didn't think all that much about it. But then Vesuvius itself started to send smoke and volcanic ash into the sky. Within a few days, tens of thousands of people were dead and two cities—Pompeii and Herculaneum—were buried under volcanic debris. Their remains were hidden from view until excavated in the 1800s and 1900s. Today, over 600,000 people live in the "red zone" around Vesuvius—the area that would be in greatest danger during a major eruption. Millions more live in the immediate vicinity, including the residents of Naples, one of the largest cities in Italy.

What was the connection between the earthquakes that occurred earlier, and the violent eruption of Vesuvius in AD 79? Vesuvius has had dozens of small eruptions since its large eruption of AD 79, but has not erupted since 1944, a longer than normal gap. When will it erupt again? How large will its eruption be? How do government officials decide when to evacuate the red zone? What will happen if they evacuate the red zone, and the volcano does not erupt for many months? These are the sorts of questions volcanologists—geologists who study volcanoes—try to answer, not only for Vesuvius, but for hundreds of volcanoes in populated areas around the globe.

Objectives

After studying this chapter and completing the exercises, you should be able to do each of the following tasks, using supporting terms and principles as necessary.

1. Relate volcanic activity to the theory of plate tectonics.
2. Explain how volcanic arcs form along convergent plate boundaries.
3. Describe the formation of the chain of volcanoes that make up the Hawaiian Islands volcanic chain.
4. Distinguish among different types of volcanic products, such as various kinds of lava flows and pyroclastic materials.
5. Distinguish among three types of volcanoes.
6. Compare the various types of igneous intrusions with each other.
7. Describe how rocks deform along a fault before an earthquake and rebound after an earthquake.
8. Distinguish among three types of seismic waves.
9. Compare and contrast the Richter scale and Mercalli scale.
10. Explain how geophysicists determine the distance between a seismometer and an earthquake's epicenter, and how they use this information to determine the location of the epicenter.
11. Contrast the characteristics of a tsunami in the deep ocean with its characteristics when it comes ashore.
12. Describe how geophysicists discovered Earth's mantle and core.
13. Compare and contrast Earth's crust, mantle, and core.

Vocabulary Terms

You should be able to define or describe each of these terms in a complete sentence or paragraph.

1. aa
2. ash
3. basement rocks
4. batholith
5. block
6. bomb
7. caldera
8. cinder
9. cinder cone
10. composite cone
11. crater
12. crust
13. dike
14. earthquake
15. elastic rebound
16. epicenter
17. fault scarp
18. fissure eruption
19. focus
20. hotspot
21. lahar
22. lapilli
23. magma chamber
24. magnitude
25. mantle
26. Mercalli intensity scale
27. Moho
28. nuée ardente
29. *P* wave
30. pahoehoe
31. peridotite
32. pillow lava
33. pipe
34. pluton
35. pyroclastic
36. pyroclastic flow
37. Richter scale
38. *S* wave
39. seismic wave
40. seismogram
41. seismometer
42. shield volcano
43. sill
44. surface wave
45. tsunami
46. volcanic arc
47. volcano
48. xenolith

7.1 Volcanoes and Plate Tectonics

Figure 7.1 shows a map of the worldwide distribution of volcanoes. Comparing this map to Figure 6.20, it is clear that there is a connection between plate tectonics and volcanic eruptions. Most volcanoes of the world are located at or near tectonic plate boundaries, especially convergent boundaries involving oceanic crust. Most prominent on this map is the "Ring of Fire," a horseshoe-shaped belt of volcanic and earthquake activity that wraps around the Pacific Ocean. The Ring of Fire, illustrated in Figure 7.2, is home to about 75% of the Earth's active volcanoes, and 80–90% of the world's earthquakes. Most of these volcanoes and earthquakes are related to the subduction of oceanic plates. A second band of volcanic and earthquake activity occurs in the Alpine-Himalayan belt. This line of mountains extends from the Alps in Europe through the Himalayas and other mountain ranges of Asia, to the volcanic regions of Southeast Asia, such as in Indonesia.

7.1.1. Volcanism at Divergent Plate Boundaries

Volcanoes occur above places where magma rises from deeper levels in Earth's lower crust and upper mantle. One place where magma rises from deeper levels is at divergent plate boundaries, discussed in Chapter 6. This magma is created from the melting of mantle rocks, and is always basaltic (mafic) in composition. Most of this magma crystallizes in the crust beneath the rift valley that runs down the cen-

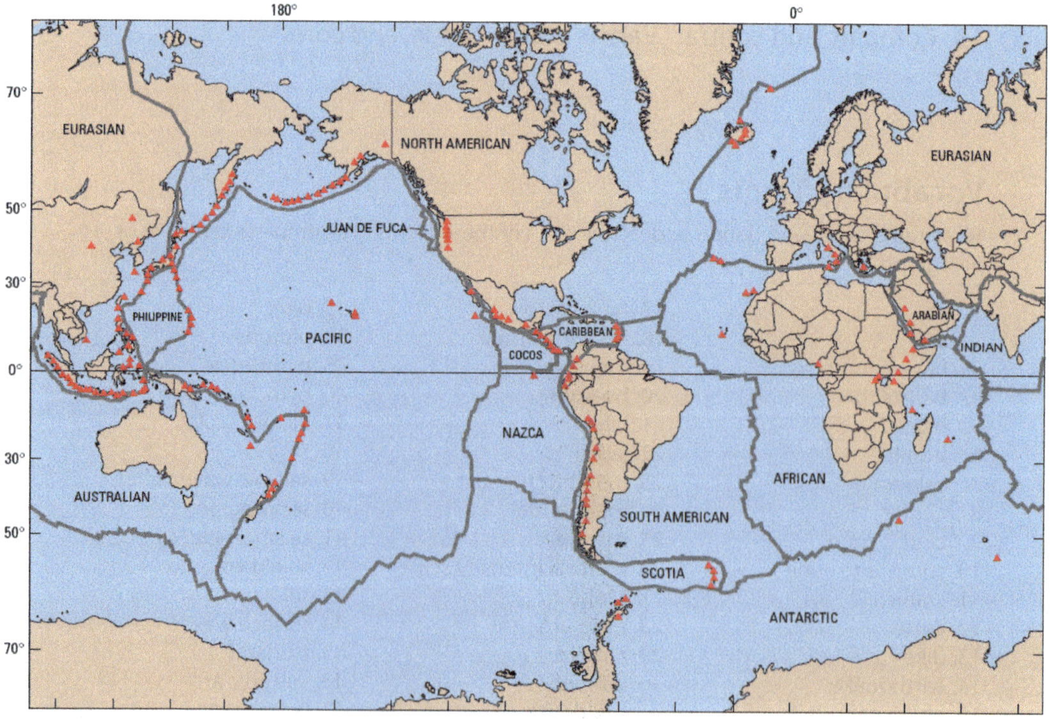

Figure 7.1. The world's major volcanoes are indicated by red triangles; tectonic plate boundaries are indicated by gray lines.

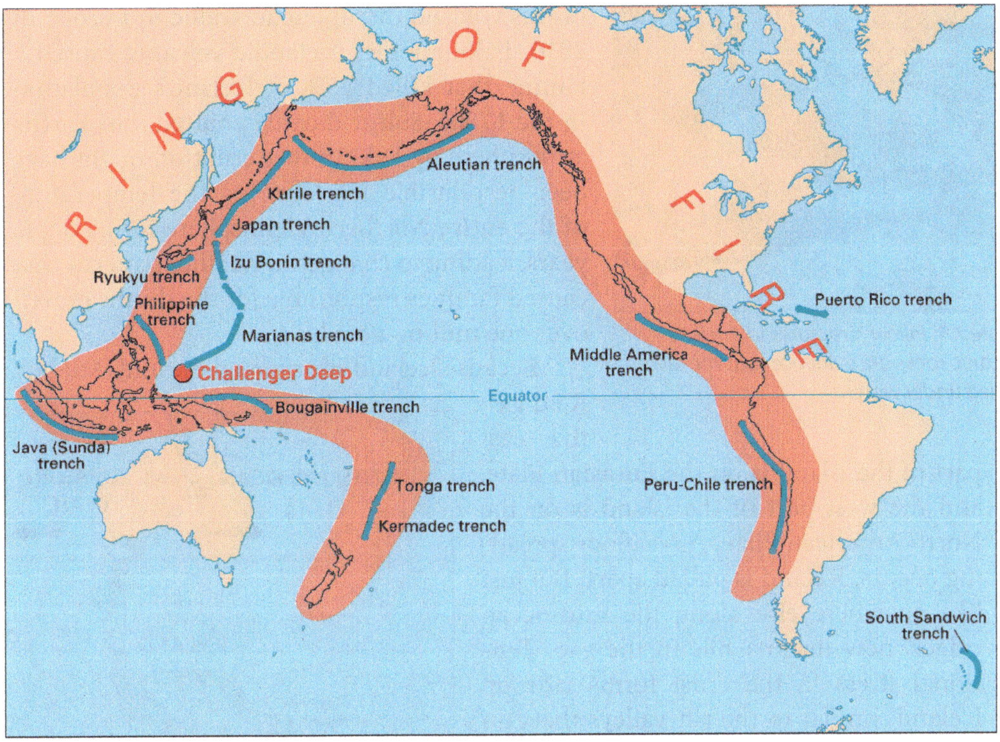

Figure 7.2. The Ring of Fire is a zone around the Pacific Ocean where the majority of Earth's volcanic eruptions and earthquakes occur. Most of these events are related to the subduction of oceanic plates. Deep-sea trenches are shown in darker blue.

ter of the mid-ocean ridges, or in the rift itself. When lava erupts underwater, such as at the mid-ocean ridges, it solidifies in distinctive pillow-shaped masses known as *pillow lava*, shown in Figure 7.3. Pillow lavas form when the outermost part of hot lava crystallizes instantly as it contacts water, forming a glassy crust on the outside of the lava. The inner part of the pillow cools more slowly, forming a rock with a fine-grained texture typical of all volcanic rocks. When examining ancient volcanic rocks on land, the presence of pillow lava, such as shown in Figure 7.4, indicates that the rocks formed under water rather than on land.

Figure 7.3. Pillow lava forms when lava flows into bodies of water. Each "pillow" is up to a meter in diameter.

For the most part, the volcanism along the mid-ocean ridges occurs in water that is thousands of meters deep. One important exception is at Iceland, an island sitting right on the Mid-Atlantic Ridge, as shown in the map of Figure 7.5. Iceland is the home to a number of volcanoes, such as Eyjafjallajökull. An eruption

Figure 7.4. Pillow structures in ancient volcanic rocks are evidence that the lava cooled under water.

of this volcano disrupted air traffic in Europe in 2010. Laki, another Icelandic volcano, caused a famine that killed 25% of Iceland's population in the 1780s. Sulfur dioxide emitted into Earth's atmosphere by Laki in that eruption might have been responsible for lowering the temperature of the entire Northern Hemisphere for a couple years, leading to changes in weather patterns that caused famines responsible for the deaths of well over one million people.

Because Iceland sits right on the mid-Atlantic Ridge, the east part of the island is on the Eurasian Plate, and the west part of the island is on the North American Plate. As seafloor spreading occurs, Iceland is pulled apart. But just like anywhere else along the mid-ocean ridges, new magma fills in the gap. Tensional stress in the crust forms rifts on Iceland, similar to the rift valleys that run along other divergent plate boundaries. One of these rifts is shown in Figure 7.6.

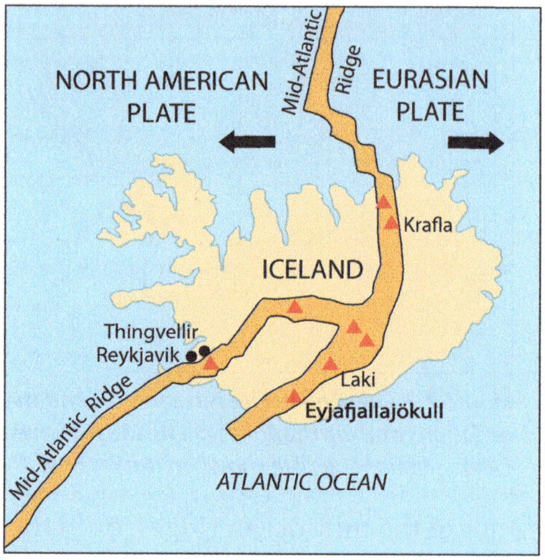

Figure 7.5. Iceland straddles the mid-Atlantic Ridge.

The East African Rift (Figure 6.22) is another place where there is a divergent plate boundary. There are a number of volcanoes in or near the rift, such as Mount Kilimanjaro and Mount Kenya, neither of which has erupted in recorded human history. Other volcanoes are more active, such as Mount Nyiragongo—a volcano with a molten lava lake in its crater (Figure 3.11).

Figure 7.6. People walking through a currently inactive rift on the Mid-Atlantic Ridge in Iceland.

7.1.2. Volcanism at Convergent Plate Boundaries

Another place where magma is created in Earth's mantle is where oceanic crust subducts into the mantle at deep-sea trenches. As oceanic plates creep deeper beneath other plates, their temperatures increase, and eventually the subducted oceanic crust begins partially to melt (remember that some minerals have higher melting points than other minerals). The molten portion gathers together, and being less dense than the surrounding rocks, begins to rise up into the crust, forming magma chambers, as illustrated in Figure 7.7. A *magma chamber* is a pool of molten or partially molten rock in Earth's crust.

Because the slab of oceanic crust subducts at an angle—typically around 45°—the magma produced by the melting of subducted crust does not form directly beneath deep-ocean trenches, but farther inland, as indicated by Figure 7.8. Magma is often generated over 100 kilometers "inland" of the deep sea trench. Most deep-sea trenches have a line of volcanoes that run parallel to the trench. This line of volcanoes is referred to as a *volcanic arc*. Many volcanic arcs are on land, such as the Cascade Range of western North America. This arc stretches from northern California to southwest British Columbia, and includes volcanoes such as Mount Shasta, Mount Hood, Mount St. Helens, and Mount Rainier. All of these volcanoes have magma chambers beneath them. The magma chambers accumulate

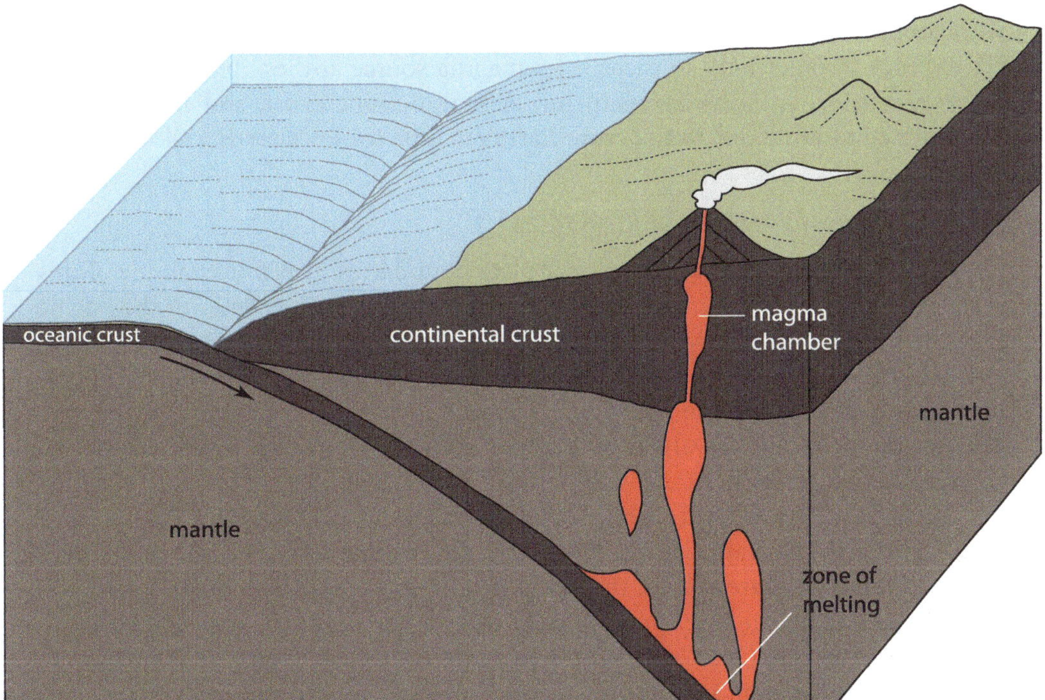

Figure 7.7. Formation of magma by melting of a subducting oceanic plate. As oceanic crust is subducted deeper into Earth's mantle, its temperature increases, and eventually it is hot enough to melt in places.

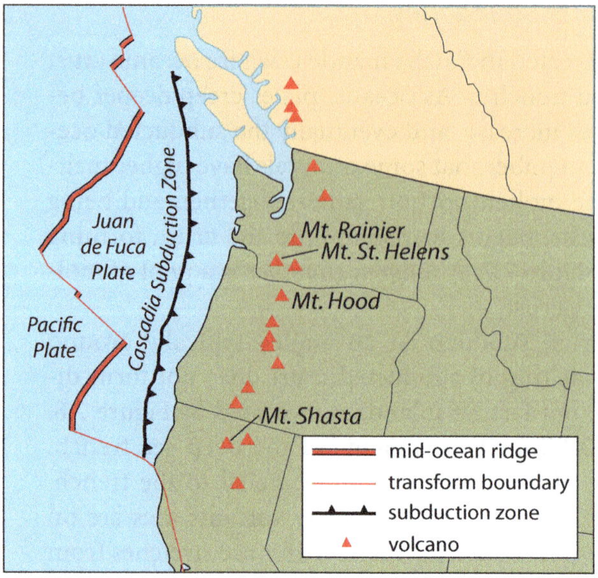

Figure 7.8. Cascade Range volcanic arc. The magma for the Cascade range volcanoes comes from the melting of the Juan de Fuca Plate. The plate partially melts as it subducts beneath the North American Plate.

supplies of magma that come up from melting oceanic crust as it is slowly subducted beneath North America.

The oceanic crust that melts to produce magma for volcanic arcs is made out of basaltic rock. But because it only partially melts as it subducts, the resulting magma has a relatively high silica content. Because of this, the volcanoes of volcanic arcs are typically made up of andesite or rhyolite rather than the basalt of mid-ocean ridges.

There is not much volcanism associated with continental-continental plate convergence, such as occurs in the Himalayas. There are two reasons for this. First, when continental crust converges with continental crust, neither gets subducted down into Earth's mantle where the crustal rocks would melt, so there is no magma source at these plate boundaries. Second, continental crust is much thicker than oceanic crust, and this makes it more difficult for magma to rise from Earth's mantle (or lower crust) all the way to the surface.

7.1.3. Volcanism Within Tectonic Plates

Not all volcanism occurs at or near plate boundaries. A prime example of this is the volcanism that occurs in the Hawaiian Islands, near the center of the Pacific Plate. A widely-accepted hypothesis is that the Hawaiian Islands sit above a *hotspot*, a region of rocks in Earth's mantle that are hotter than the typical mantle temperature. This heat causes some rocks in the mantle to melt, creating basaltic magma. The magma then rises toward the crust, and some of it makes it all the way to the surface to erupt at volcanoes.

The Hawaiian hotspot is currently beneath the island of Hawaii, the easternmost and largest of the Hawaiian Island chain, the main part of which is shown in Figure 7.9. As the Pacific Plate moves, the hotspot stays in one place. Moving northwest from Hawaii, the islands get progressively older. Maui was over the hotspot about 1.3 million years ago (Ma), and Oahu was over the hotspot about 3.4 Ma. Volcanic activity is very frequent on Hawaii; Kilauea has been erupting continuously since 1983. The last time Haleakala had an eruption was in 1790. There are small volcanoes still on Oahu, but they have not erupted in human memory. There is no longer any active volcanism on Kauai or Niihau, and the volcanoes on these islands are deeply eroded by streams. Figure 7.10 shows how the Hawaiian Islands are moving over the Hawaiian hotspot.

Figure 7.9. The main part of the Hawaiian Islands. Each island was created by volcanoes, but the most active volcanoes (red triangles) are all at the east end of the chain. Loihi is the youngest volcano in the chain, and its summit is still about 3,000 meters beneath the surface of the ocean.

Northwest of Hawaii, there is a chain of small islands and atolls all in a row, including Midway Island, the site of an important battle in WW II. Even farther to the northwest, there is a long chain of seamounts, called the Emperor Seamounts, trailing all the way to the edge of the Pacific Ocean near Russia. These seamounts, shown in Figure 7.11, were once large volcanic islands like Hawaii, but they subsided deep into the ocean once they no longer had a magma source. The oldest volcanoes in the chain, off the coast of Russia, are about 85 million years old. The newest volcano in the chain is Loihi, a seamount off the southeastern coast of Hawaii that could eventually form the newest island in the Hawaiian Islands. Loihi appears on both Figure 7.9 and 7.10.

Another volcanic area that is in the interior of a tectonic plate, rather than near a plate boundary, is found in Yellowstone National Park in Wyoming. The Yellowstone Caldera[1] is one of the largest volcanoes in the world, and is underlain by a massive magma chamber, illustrated back in Figure 5.3. This magma chamber is about 80 kilometers (50 miles) long and 20 kilometers (12 miles) wide, with a total

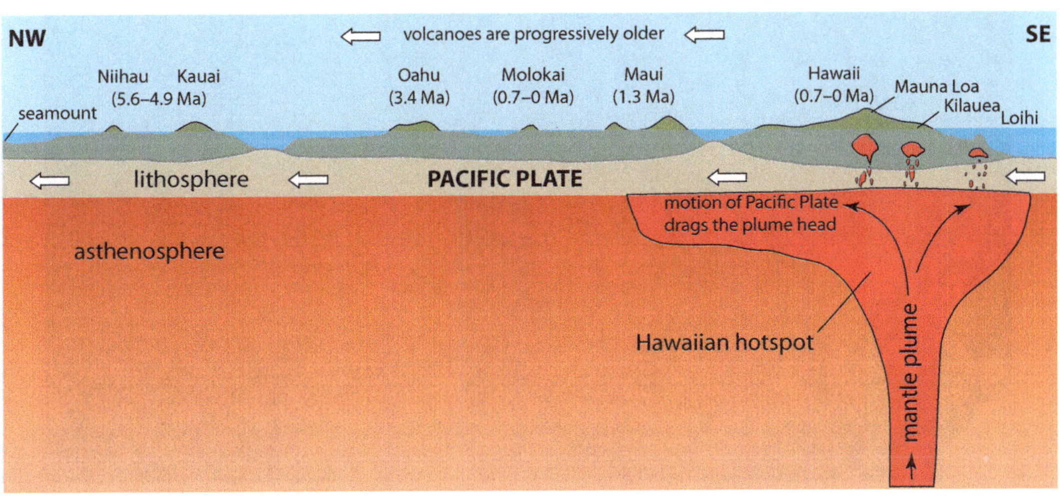

Figure 7.10. The Hawaiian hotspot is presently under the big island of Hawaii. As the Pacific Plate slowly moves over the hotspot, volcanoes on older islands become extinct and new islands form from new volcanoes.

1 The term "caldera" will be defined later in the chapter.

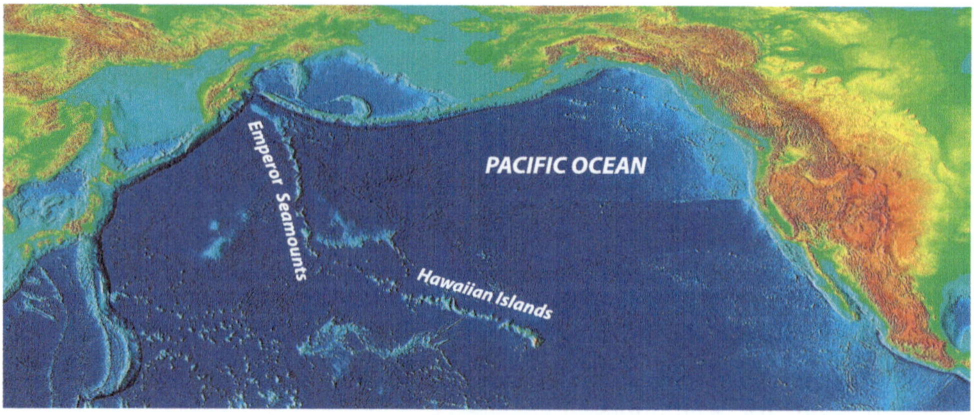

Figure 7.11. The Hawaiian Islands are part of a longer chain of extinct volcanoes that stretches to the far northwestern Pacific Ocean, and includes the Emperor Seamounts. Each seamount was once a large volcanic island sitting over the Hawaiian hotspot.

volume of about 4,000 km^3. Like most magma, this magma is a slushy mix of molten rock and solid mineral crystals. The North American plate is moving over the Yellowstone hotspot at a rate of about 3 centimeters per year, and there are rocks formed by older calderas in a line running across southern Idaho, illustrated in the map of Figure 7.12.

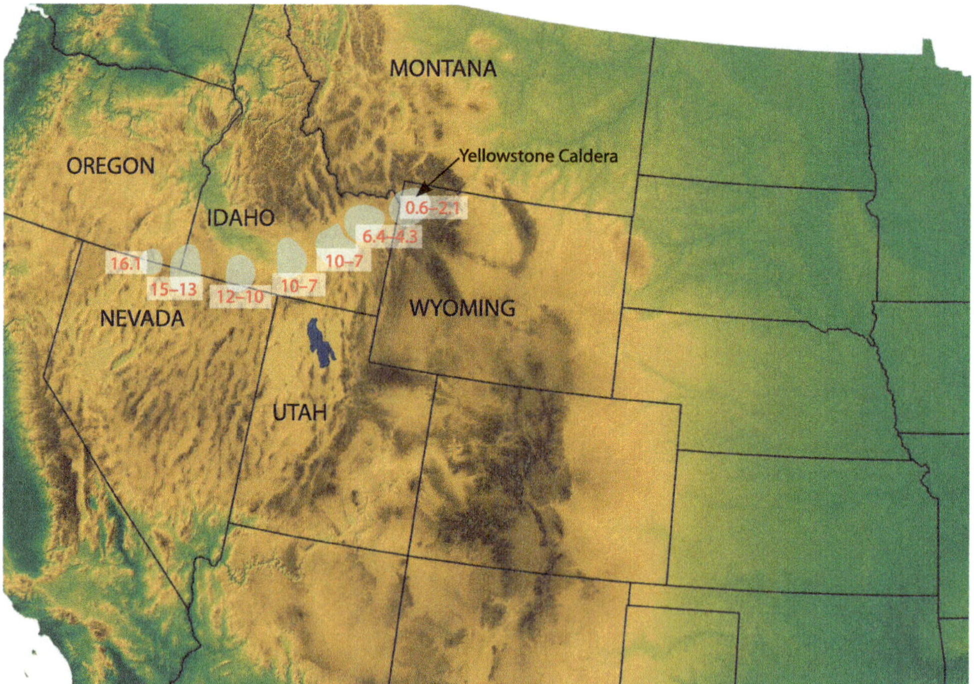

Figure 7.12. The current Yellowstone Caldera formed over the period of 0.6 to 2.1 million years ago. Rocks from older caldera eruptions curve across southern Idaho. The oldest known eruptions from the Yellowstone hotspot occurred along the present Oregon-Nevada border about 16.1 million years ago.

Learning Check 7.1

1. What is the Ring of Fire?
2. Describe the distribution of volcanoes on Earth, and explain why most volcanoes occur where they do.
3. How does pillow lava form? If a geologist finds pillow lava in an ancient deposit, what can the geologist conclude about the formation of the lava?
4. How does subduction of oceanic crust at a deep-sea trench lead to the formation of magma?
5. Explain how the geologic history of the Hawaiian Islands is related to plate tectonics, even though Hawaii is far from a tectonic plate boundary.

7.2 Volcanoes

Volcanic eruptions are among the most dramatic of geological phenomena. Many geological processes, such as plate tectonics, weathering, and growth of coral reefs, occur at very slow rates. Volcanic eruptions, on the other hand, can effect a tremendous amount of change in a very short time. We often think of volcanoes as being destructive—and they can be—but they also seem to be an integral part of the processes God used to form the Earth. Generally, they are only considered to be "bad" when they cause harm. However, volcanoes are also beneficial to humans because they help create some of the richest soils on the planet.

7.2.1. Volcanic Materials

Back in Chapter 5, you read about magmas and lavas with various compositions, such as mafic magma, which produces rocks such as basalt and gabbro; intermediate magma, which produces andesite and diorite; and felsic magma, which produces rhyolite and granite. Mafic magma, with its relatively low silica content, has a low viscosity, meaning that it flows easily. Rhyolite has high viscosity and does not flow very well at all. Also, mafic magma does not contain a lot of dissolved gases, while rhyolitic magma is typically gas-rich. The most common gases present in magma are water vapor, carbon dioxide, and sulfur dioxide. Viscosity and gas content affect the nature of volcanic

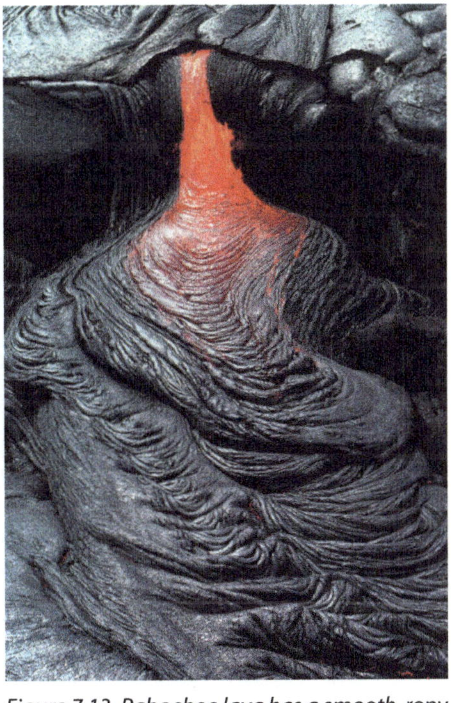

Figure 7.13. Pahoehoe lava has a smooth, ropy surface.

Figure 7.14. Aa is a jagged-surfaced basaltic lava flow. These scientists are pacing distances on the road to measure the rate of advance of the lava.

eruptions and the types of materials produced in those eruptions.

Many volcanic eruptions involve the production of lava flows. Because of its low viscosity, basaltic lava can flow for long distances: the longest known flows are hundreds of kilometers long! Basaltic magma has been clocked moving at rates of up to 30 kilometers per hour (20 miles per hour), but speeds of around 5 to 100 meters per hour are much more typical. As they solidify, many basaltic lava flows form a smooth skin that wrinkles as the lava continues to move downhill. This smooth-surfaced volcanic rock, pictured in Figure 7.13, is called *pahoehoe*.[2] Some basaltic lava is more viscous, and breaks up into sharp fragments as the lava advances. These rocks, shown in Figure 7.14, are referred to as *aa*.[3] Pahoehoe and aa are both Hawaiian words, and both types of rock are abundant there. One can walk on a cooled pahoehoe surface in bare feet, but walking on aa rocks will tear up the thick soles of hiking boots. Aa flows advance at rates as slowly as one meter per day, but can relentlessly advance for months at a time.

Figure 7.15. The Mount St. Helens eruption in Washington State in 1980 was an explosive eruption that killed 57 people, caused one billion dollars of damage, and deposited wind-blown ash in 11 U.S. states and 5 Canadian provinces. Schools had to be closed as far as 1,000 km (600 miles) downwind due to accumulations of volcanic ash.

Lava flows also form from andesite and rhyolite. Because of their high viscosity, these flows usually do not move rapidly, but they can still be quite destructive over time.

2 Pronounced PAH-hoy-hoy.

3 Pronounced AH-ah.

The gas content of magma is a key factor in determining whether a volcanic eruption will be explosive or non-explosive. Eruptions of intermediate and felsic lava, such as typically found in volcanic arcs or within continents, are generally explosive. As magma moves upward, gases come out of solution, like gas bubbles released when a can of soft drink is opened. The forces produced by these expanding gases can be enough to propel volcanic materials thousands or even tens of thousands of meters into Earth's atmosphere. In extreme cases, such as the 1980 eruption of Mount St. Helens in Washington depicted in Figure 7.15, an explosive eruption destroys a significant portion of a volcanic mountain. In contrast, some volcanic eruptions are relatively gentle. Kilauea in Hawaii has been erupting continuously since 1983, mostly producing fluid lava flows, many of which run anywhere between 10 and 25 kilometers to the Pacific Ocean. Tourists, such as those shown in Figure 7.16, safely watch these eruptions from a distance.

Figure 7.16. Many basaltic volcano eruptions are non-explosive and can be safely viewed from a short distance.

As with most substances, lava flows contract when they cool down. This sometimes causes the rocks to fracture into a regular, polygonal pattern called *columnar joints*, shown in Figure 7.17.

Pyroclastic[4] materials are those expelled from volcanoes and into the air, such as the material shown in Figure 7.18. These materials range in size from boulders weighing more than a ton, to microscopic dust particles. Volcanic *ash* is silt and sand-sized particles that can be carried in the atmosphere thousands of kilometers downwind from a volcano. Ash is formed during an explosive eruption when gases rapidly expand, making a frothy mix of gas

Figure 7.17. Columnar joints in basalt at Devil's Postpile National Monument in California.

4 *Pyroclastic* comes from Greek words meaning "fire" and "fragment."

Figure 7.18. Ash from the 2010 eruption of Eyjafjallajökull in Iceland caused the cancellation of about 100,000 airline flights in Europe, affecting ten million passengers.

bubbles and volcanic glass. The column of material coming out of Mount St. Helens in Figure 7.15 was composed largely of steam and ash. Ash is like pumice, shown in Figure 5.9, except that it has been shattered into much smaller pieces.

Medium-sized pyroclastic particles are called *cinders* if they are pea-sized, and *lapilli* if walnut-sized. Larger pyroclastic particles are called *blocks* if they are formed of hardened lava and *bombs* if they were ejected from the volcano as molten rock. Bombs, such as the one shown in Figure 7.19, often have a streamlined shape like an American football. The shape forms as the semi-molten bomb flies through the air. Blocks and bombs typically fall on the slopes of the volcano, though they can be propelled farther away if the force of the escaping gases is sufficient.

Figure 7.19. Volcanic bombs are shaped as liquid lava hurls through the atmosphere.

The pyroclastic materials discussed so far are all individual particles ejected during an eruption. A *pyroclastic flow* is a much more massive event. A pyroclastic flow is a mass of hot, glowing pyroclastic material ejected from a volcano. Pyroclastic flows can have enormous volumes, and travel across landscapes at speeds greater than 100 kilometers per hour (60 miles per hour). After coming to a rest, these hot materials fuse together to form a solid rock similar to a lava flow.

7.2.2. Types of Volcanoes

A *volcano* is a mountain built by accumulations of volcanic materials, such as lava and pyroclastic materials. Most volcanoes have a steep-walled depression at the summit called a *crater*. Beneath the crater is a *pipe*, which connects the crater to the magma chamber at depth. An unusually large crater with a diameter greater than 1 km is called a *caldera*.

Fluid lava, such as found at Hawaii, forms volcanoes that have a broad dome shape. These volcanoes are called *shield volcanoes*, because they look like a warrior's shield lying flat on the ground. Figure 7.20 shows Mauna Loa, a shield volcano in Hawaii. Because basalt lava usually flows easily, shield volcanoes spread out over large areas and have gentle slopes. Shield volcanoes are the largest mountains on Earth, both by volume and by measurements from their base to their summits. Mauna Loa has a base at a depth of 5,000 meters below sea level, and a summit at over 4,000 meters above sea level, giving it a relief of over 9,000 meters (about 5.5 miles). Measured from its base to its summit, Mauna Loa is taller than Mount Everest.

Figure 7.20. Mauna Loa on Hawaii is a shield volcano. Mauna Loa is Earth's largest volcano in terms of volume, area of its base, and relief measured from its base to summit. Shield volcanoes have gentle slopes because basaltic lava flows for long distances. Mauna Kea, another shield volcano, can be seen in the distance.

Shield volcanoes also exist on Mars and Venus. The tallest mountain in the entire solar system is Olympus Mons, a shield volcano on Mars, shown in Figure 7.21. Olympus Mons is 23 kilometers (14 miles) high, and has a base 600 kilometers (370 miles) in diameter. This volcano is roughly the size of the state of Arizona.

Many of the volcanoes in volcanic arcs—such as the major volcanoes of the Cascade Range and other volcanic arcs of the Ring of Fire—are *composite*

Figure 7.21. Olympus Mons is a shield volcano on Mars. It is the tallest and most massive mountain in the solar system.

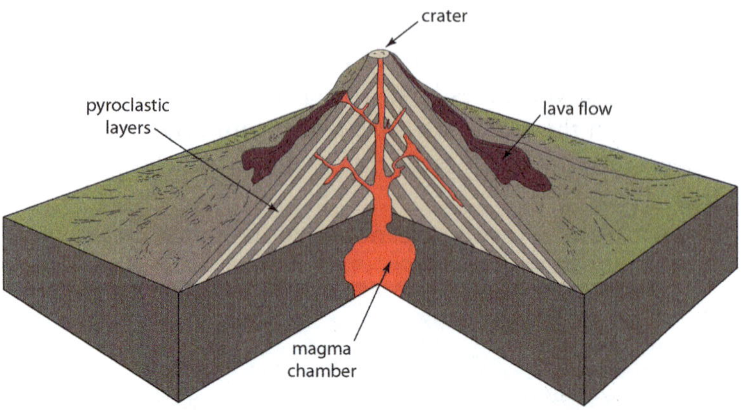

Figure 7.22. The internal structure of a composite cone, composed of alternating layers of lava flows and pyroclastic flows.

cones. A composite cone[5] is a volcano that has alternating layers of lava flows and pyroclastic material, as illustrated in Figure 7.22. Because they are made up from less fluid materials, composite cones have steeper slopes than shield volcanoes. Composite-cone volcanoes often produce violent eruptions and present a considerable risk to human populations that live near them. Most of the well-known volcanoes of the world, such as Vesuvius in Italy, Mount Fuji in Japan, and the main volcanoes of the Cascade Range in the United States, are composite cones. Figure 7.23 shows Mount Hood, a composite cone in the Cascade Range.

Figure 7.23. Mount Hood in Oregon has the typical cone shape and steep slopes of a composite cone.

One hazard from composite cones is a pyroclastic flow called a *nuée ardente*.[6] A nuée ardente is

Figure 7.24. The cloud moving down the flanks of this volcano is a nuée ardente, a very hot, rapidly moving mixture of steam and pyroclastic material. The eruption occurred in 1984 at Mayon Volcano in the Philippines. Officials evacuated 70,000 people from the base of the volcano. Happily, no one was killed.

5 Composite volcanoes are also called *stratovolcanoes*.

6 Pronounced noo-EH ar-DAWNT. *Nuée ardente* means "burning cloud" in French.

Figure 7.25. The city of Armero, Colombia used to be in the center of the area represented in this photograph, but was almost completely buried by a lahar in 1985. The death toll was 23,000.

a mixture of pyroclastic materials, very hot steam, and other hot gases, and it moves rapidly down the slope of a volcano with very little friction. These glowing clouds, such as the one shown in Figure 7.24, travel at speeds approaching 200 kilometers per hour (125 miles per hour). In 1902, a nuée ardente from Mount Pelée, on the Caribbean island of Martinique, destroyed the port town of St. Pierre, killing 28,000 people virtually instantly. There were only three survivors in the town, a shoemaker, a little girl, and a convicted prisoner locked in a prison dungeon. Most of the people on ships in the harbor were killed as well. The nuée ardente destroyed St. Pierre, but actually left behind only a thin deposit of pyroclastic material.

Another hazard associated with composite cones is mudflows called *lahars*. A lahar is a thick, heavy mixture of water and pyroclastic material that flows down the slope of a volcano. The water can come from heavy rainfall occurring during or after an eruption or from melting of snow and ice on the slopes of the volcano. In 1985, lahars from a relatively small eruption of the volcano Nevada del Ruiz, in Colombia, killed about 23,000 in and around the town of Armero, about 100 kilometers from the volcano. In Figure 7.25, you can see that Armero was almost completely buried by the lahar. The possibility of lahars is considered to be the greatest risk if glacier-covered Mt. Rainier erupts in Washington. The lahars could easily reach developed areas in lowlands around Seattle and Tacoma. Several towns west of Mt. Rainier now have lahar early-warning systems in place.

Figure 7.26. Parícutin is a cinder cone in Mexico. It formed in the 1940s.

A third type of volcano is a *cinder cone*. Cinder cones are small volcanoes built from loose pyroclastic material such as cinders and lapilli.

Cinder cones are much smaller than either composite cones or shield volcanoes. They are typically less than 300 meters (1,000 feet) tall, and often form in groups. Geologists were able to observe the growth of a cinder cone from beginning to end when Parícutin volcano formed in central Mexico

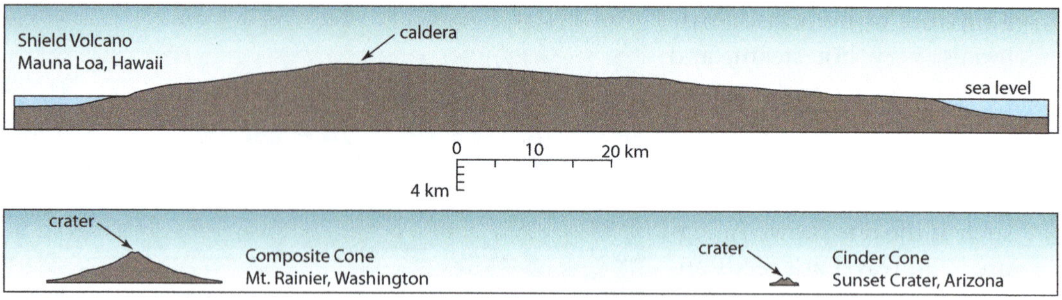

Figure 7.27. Profiles and size comparison of the three main types of volcanoes.

starting in 1943. A farmer observed sulfurous smoke coming out of a hole in his cornfield, and within a week the cone built to a height of over 100 meters. Within two years, the cone, pictured in Figure 7.26, built to a height of over 400 meters and produced some aa lava flows as well, destroying two villages. The eruptions ceased in 1952. Parícutin will probably never erupt again; future eruptions in the area will probably create new cinder cones. There are numerous cinder cones in the western United States, such as Capulin Volcano in New Mexico and Sunset Crater in Arizona, a cinder cone that formed about 1,000 years ago. The sizes of the three types of volcanoes—shield, composite cone, and cinder cone—are compared in Figure 7.27.

7.2.3. Caldera Eruptions

As mentioned already, a caldera is a volcanic crater that is at least one kilometer in diameter. Calderas form at the end of volcanic eruptions, and form when the summit of a volcano collapses into a partially empty magma chamber. Some of the shield volcanoes in Hawaii have calderas at their summits.

One place that illustrates caldera formation is Crater Lake in Oregon, shown in Figure 7.28. Crater Lake is the deepest lake in the United States, and occupies a caldera formed from what geologists have named Mt. Mazama. Mt. Mazama was largely destroyed by a major eruption about 7,000 years ago. Figure 7.29 shows the sequence of events that led to the formation of this caldera. In this eruption, Mt. Mazama ejected more than 50 cubic kilometers of pyroclastic material, leaving the magma chamber partially empty. The roof of the magma chamber collapsed, creating the caldera. There were a few smaller eruptions after the formation of the caldera, leading to the formation

Figure 7.28. Crater Lake in Oregon is a caldera formed by the eruption and collapse of Mt. Mazama in about 5000 B.C. The caldera is about 10 kilometers (6 miles) in diameter. Wizard Island, in the foreground, is a cinder cone that formed sometime after the eruption of Mt. Mazama.

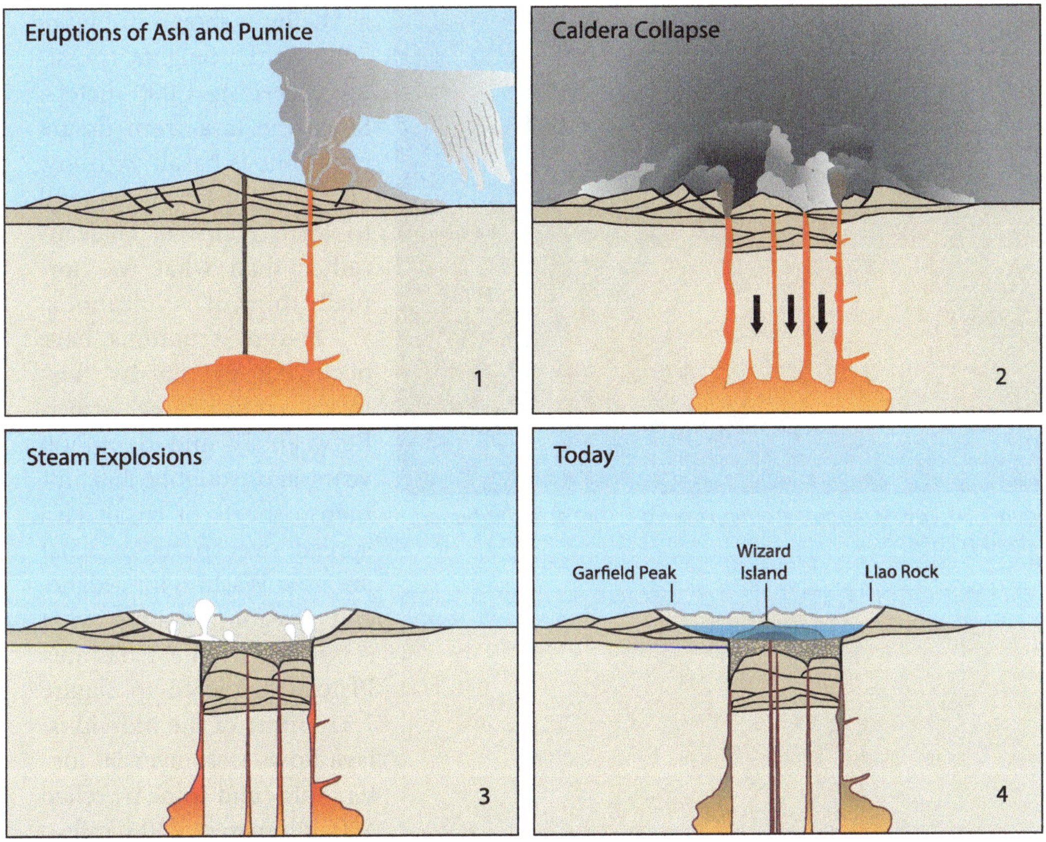

Figure 7.29. The sequence of events that led to the formation of Crater Lake. When the magma chamber partially emptied, its roof collapsed, forming a caldera.

of Wizard Island—a cinder cone—as well as some smaller cones that are now under water. The caldera has now partially filled with rainwater.

Some calderas are so large that they were not noticed until geologists did careful mapping of volcanic areas. One example of this is the Yellowstone Caldera. The Yellowstone Caldera has had three "supervolcano" eruptions in its history, roughly 2.1, 1.3, and 0.6 million years ago. The largest of these eruptions ejected about 2,000 km^3 of pyroclastic material and blanketed much of the United States with thick deposits of volcanic ash. The caldera was difficult to discover because it has mostly filled with smaller, more recent lava flows.

7.2.4. Fissure Eruptions

Not all volcanic eruptions fit into the categories of shield volcano, composite cone, and cinder cone. The eruption of Laki in Iceland in 1783–1784, mentioned in Section 7.1, is an example of a *fissure eruption*. Rather than erupting from a central crater on a volcano, in a fissure eruption lava erupts from long cracks in Earth's crust called *fissures*. At Laki, a rift 25 kilometers long erupted more than 15 km^3 of very fluid basalt. This lava spread out over an area of about 600 square kilometers (230 square miles). At times, the lava erupted as enormous lava fountains—such

Figure 7.30. Lava fountains erupting from a fissure at Bárðarbunga, Iceland in 2014.

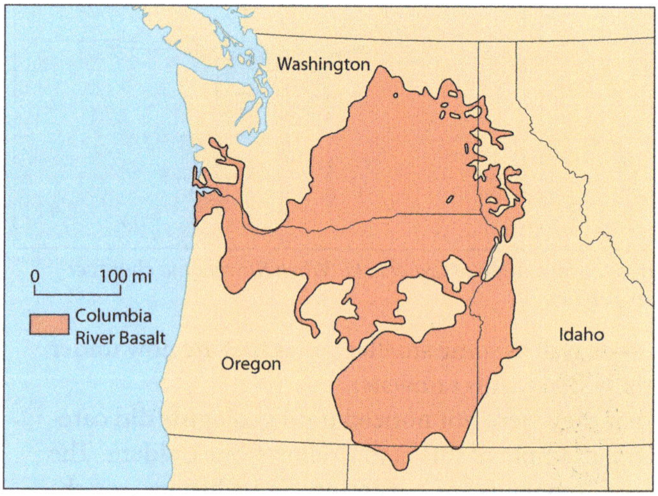

Figure 7.31. The dark area shows the extent of the Columbia Plateau lava flows in the northwestern United States.

as the ones shown in Figure 7.30—with heights possibly exceeding 1,000 meters. Since the lava from fissure eruptions is basalt with low viscosity, the lava flows tend to form plains or plateaus rather than what we normally think of as volcanoes.

Fissure eruptions have occurred in Earth's history at much larger scales. Between 17 and 6 million years ago, roughly 350 immense sheets of basalt lava spread over parts of what are now Washington, Idaho, and Oregon, creating what is known as the Columbia Plateau, depicted in Figure 7.31. Some of the individual lava flows were over 50 meters thick and some travelled 500 kilometers (300 miles) from southeastern Washington to the Pacific Ocean in less than a week. The overall thickness of the basaltic lava flows on the Columbia Plateau is close to 2 kilometers (1.6 miles).

Learning Check 7.2

1. Compare and contrast pahoehoe and aa lava flows.
2. Compare and contrast lava flows, pyroclastic flows, and lahars.
3. Explain why basaltic eruptions form shield volcanoes, while andesitic eruptions form composite cones.
4. Describe how a caldera forms, and explain how a caldera is different from an ordinary volcanic crater.
5. What type of landscape is formed by lava flows from a fissure eruption?

7.3 Igneous Intrusions

Figure 7.32. A xenolith is a rock fragment that has become incorporated into magma. The black piece of gabbro broke off the walls of a magma chamber to become surrounded by light-colored granite.

The mechanism by which magma rises in the crust is an area of controversy and ongoing research. Magma rises from deep in Earth's crust because molten rock has a lower density than solid rock. Many scientists believe that at very deep levels, rocks are flexible enough that surrounding rocks can probably deform around blobs of rising magma. In much of the crust, however, this probably cannot occur. Magma also ascends by partially melting surrounding rocks, or by breaking off pieces of the surrounding rocks. It is fairly common in some places to find pieces of the surrounding crustal rocks within igneous rocks. These unmelted pieces of surrounding rocks, such as the one picture in Figure 7.32, are called *xenoliths*. One of the primary means by which magma rises through the crust may be through fractures.

Most magma never reaches Earth's surface, but crystallizes at some depth as a coarse-grained igneous rock such as granite, diorite, or gabbro. An intrusive igneous rock body is called a *pluton*.[7] Plutons form within the crust, but are often uplifted after forming and exposed through weathering and erosion.

The largest plutons are called *batholiths*, typically composed of granite or diorite. The minimum size of a batholith is 100 km^2 (about 40 square miles). Most batholiths are probably the crystallized remains of magma chambers that existed beneath volcanic arcs. The volcanoes in many of these arcs have long-since eroded away. Some large batholiths, such as the Sierra Nevada Batholith in California pictured in Figure 7.33, are hundreds of kilometers long and have volumes measured in tens of thousands of cubic kilometers. Most batholiths probably did not form in a

Figure 7.33. The rocks that make up the core of the Sierra Nevada Range in California are granitic rocks of the Sierra Nevada Batholith.

7 *Plutons* are named after Pluto, the god of the underworld in classical Greek mythology.

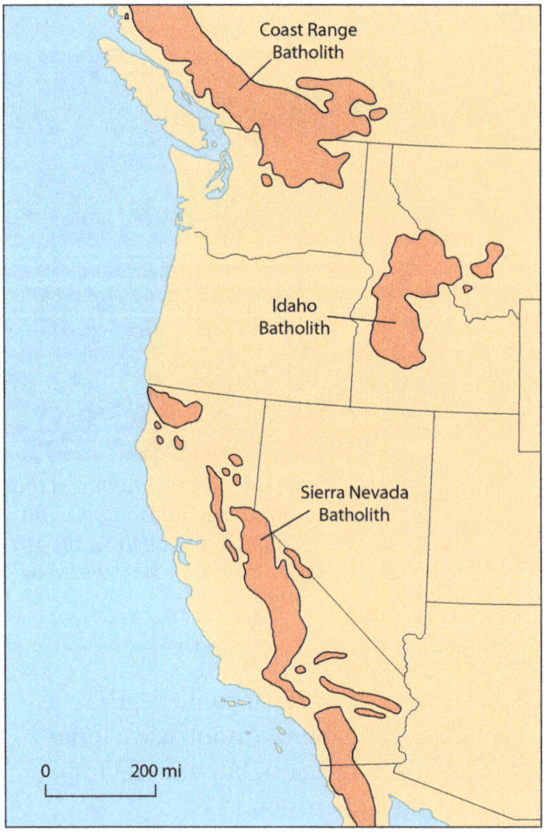

Figure 7.34. There are several large batholiths in western North America. These formed as magma chambers, but have been uplifted and uncovered through weathering and erosion.

single intrusion event; instead, there is evidence that large batholiths are actually composed of dozens or even hundreds of smaller plutons that formed at different times. Figure 7.34 is a map showing the location of some large batholiths in western North America.

Batholiths are massive bodies, having widths, lengths, and depths measured in kilometers. A *dike* is an intrusion where magma pushed into a fracture in Earth's crust. Dikes, such as the one shown in Figure 7.35, form in cracks within igneous, metamorphic, and sedimentary rocks, and range in thickness from a few centimeters to hundreds of meters. When there is layering present in the original rocks, a dike cuts across that layering. A *sill* is similar to a dike, except that sills intrude parallel to layers of rock, as illustrated in Figure 7.36. Sills usually form at shallow depths in Earth's crust, and "cook" the surrounding sedimentary rocks, making a narrow band of contact metamorphism.

Because they usually form at considerable depth within Earth's crust,

Figure 7.35. A dark dike cutting through horizontal layers of sedimentary rocks in Arizona.

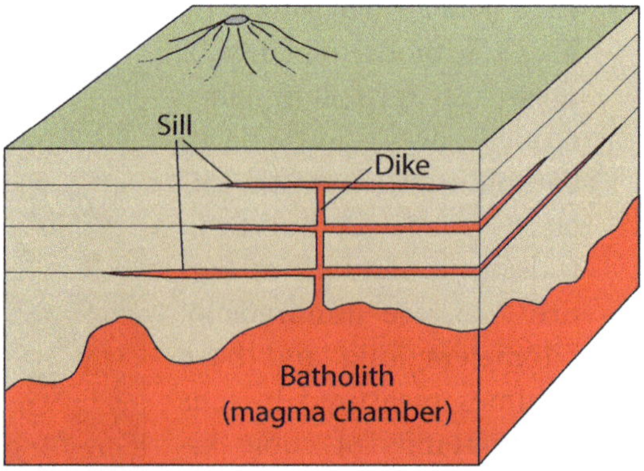

Figure 7.36. Most igneous rocks crystallize beneath Earth's surface in magma chambers, which cool to become batholiths. Dikes are planar igneous bodies which cut through rock, and sills are planar igneous bodies which squeeze in between layers of rock.

most plutons are composed of coarse-grained igneous rocks such as granite and diorite. When dikes and sills form at shallow depths, where the surrounding rocks are relatively cool, they can have a porphyritic or fine-grained texture.

The formation of most plutons can be related to the theory of plate tectonics. Most volcanoes form inland from zones of tectonic plate subduction. Because many plutons are associated with magma chambers beneath volcanoes, plutons are also found inland from subduction zones.

Learning Check 7.3

1. What is a pluton? What kinds of rocks are they made out of?
2. How are dikes and sills similar to each other? How are they different from each other?
3. Sometimes the grain size of minerals in a sill is larger in the middle of the sill than it is along the sill margins. Why do you think this happens?

7.4 Earthquakes

An *earthquake* is a violent shaking of Earth's crust caused by a rapid release of energy. Earthquakes are usually caused by sudden movement along faults (Section 6.5.2), but they are also caused by events such as volcanic eruptions, landslides, and meteorite impacts. Earthquakes are among the most destructive geologic events, and are capable of killing thousands—or even hundreds of thousands—of people in a very short time period. The study of earthquakes is called *seismology*.

7.4.1. Elastic Deformation

When rocks are subjected to stress from the forces of compression or tension, they change their shape. Low amounts of stress cause elastic deformation. This means that the rocks bend or otherwise deform, but will return to their original shapes if the stress is removed. Elastic deformation is what happens to a rubber band when it is stretched. When the tension is removed, the rubber band returns to its original shape. There are limits to how much a rock can be deformed; if those limits are exceeded the rock will break. When this happens to massive blocks of Earth's crust, an earthquake occurs. The same thing happens to familiar objects, such as a wooden pencil. You can bend a pencil a little bit, and if you let go, it will be restored to its original shape. If the pencil is bent too far, it will break. But after the pencil breaks, the two halves bounce back to their original straight shape. This is called *elastic rebound*, and the same thing happens to blocks of rock after an earthquake, as illustrated in Figure 7.37. In the illustration, an imaginary fence is built across the San Andreas fault. As the Pacific and North American plates move past each other slowly, the fence deforms. When an earthquake occurs, the fence snaps and each piece resumes its original straight shape.

Elastic deformation can be measured in the field. Scientists have placed lasers along some faults, such as the San Andreas Fault in California, in order to mea-

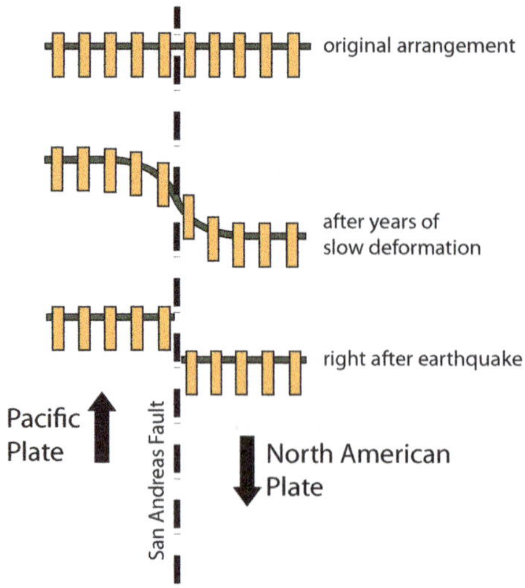

Figure 7.37. Illustrated by an imaginary fence built across the San Andreas fault, rocks on either side of a fault will be elastically deformed by stress until a breaking point is reached. This deformation can be observed and measured.

sure how much rocks are deforming along the fault line. In places, there is movement of 2–5 centimeters per year on one side of the fault relative to the other. That does not sound like much, but over a period of a century, this builds up to a few meters of deformation of massive blocks of rocks. When the limit of deformation is reached, there will be sudden movement along the fault: an earthquake.

7.4.2. Faults

Not all faults respond to stress in the same way. When stress is applied to some faults, the blocks on opposite sides of the fault glide past each other rather easily, and little elastic deformation occurs. In this case, no significant earthquakes occur. Other faults stick a little bit, allowing some elastic deformation of the rocks to occur on either side of the fault. This stress is then relieved by frequent small earthquakes. Still other faults are very sticky, causing a tremendous buildup of deformation due to elastic stress over a period of many decades or even several centuries. When these faults rupture, the result is a catastrophic earthquake. There are typically around twenty major earthquakes in Earth's crust per year, though many of these occur in remote places and do little significant damage.

Earthquakes can be caused by any of the kinds of faults mentioned in Chapter 6: normal faults, reverse faults, strike-slip faults, or thrust faults. The San Andreas Fault is an example of a strike-slip fault. An example of horizontal displacement along the San Andreas Fault is shown in Figure 7.38.

Figure 7.38. The 1906 San Francisco earthquake was caused by horizontal movement along the strike-slip San Andreas Fault. Compare this photograph to Figure 7.37.

Many mountain ranges in the western United States were uplifted by normal faults. When an earthquake occurs along a normal fault, the displacement can be seen on the ground surface as a *fault scarp*, a small cliff caused by vertical movement along a fault. Figure 7.39 shows a fault scarp. Fault scarps are most often seen in arid regions because in humid climates they begin to weather and erode soon after they are formed.

The world map in Figure 7.40 (next page) shows that most earthquakes occur close to plate boundaries. A majority of the world's major earthquakes occur around the margins of the Pacific Ocean, in places like Chile, Peru, California, Alaska, and Japan. The distribution of major earthquakes around the world is very similar to the distribution of volcanoes, illustrated in Figure 7.1.

Figure 7.39. This fault scarp was formed during an earthquake near Yellowstone National Park in 1959.

7.4.3. Seismic Waves

The vibrations in Earth's crust during an earthquake are caused by waves called *seismic waves.* Every earthquake creates three types of seismic waves: *P waves*, *S waves*, and *surface waves*, illustrated and summarized in Table 7.1. P waves (or primary waves) cause particles in Earth's crust to move back and forth in the same direction that the wave is travelling. In other words, a P wave involves compression and expansion of the material it passes through. P waves are essentially sound waves that travel through Earth's interior rather than through the atmosphere. P waves have the fastest velocity of the seismic waves, so they are the first waves to arrive at a location on Earth's surface.

P Wave	S Wave	Surface Wave
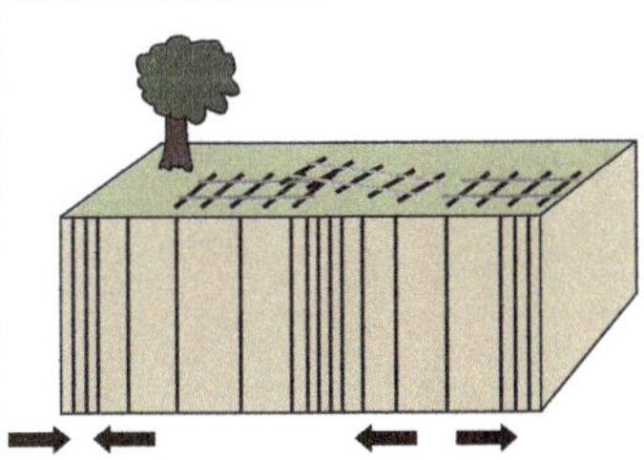	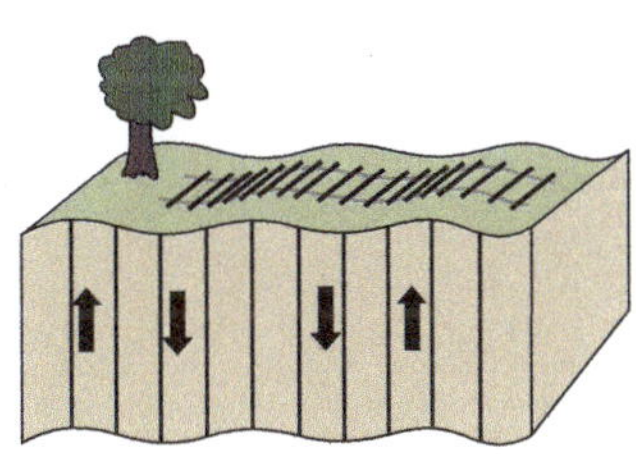	
Compression and expansion of material in direction wave is moving	Back and forth movement perpendicular to direction wave is moving	Surface moves both up and down and back and forth
fastest—6 km/s in granite near surface	intermediate speed—3.6 km/s in granite near surface	slowest—3.2 km/s in granite near surface
weakest	intermediate intensity	strongest shaking

Table 7.1. Characteristics of Seismic Waves.

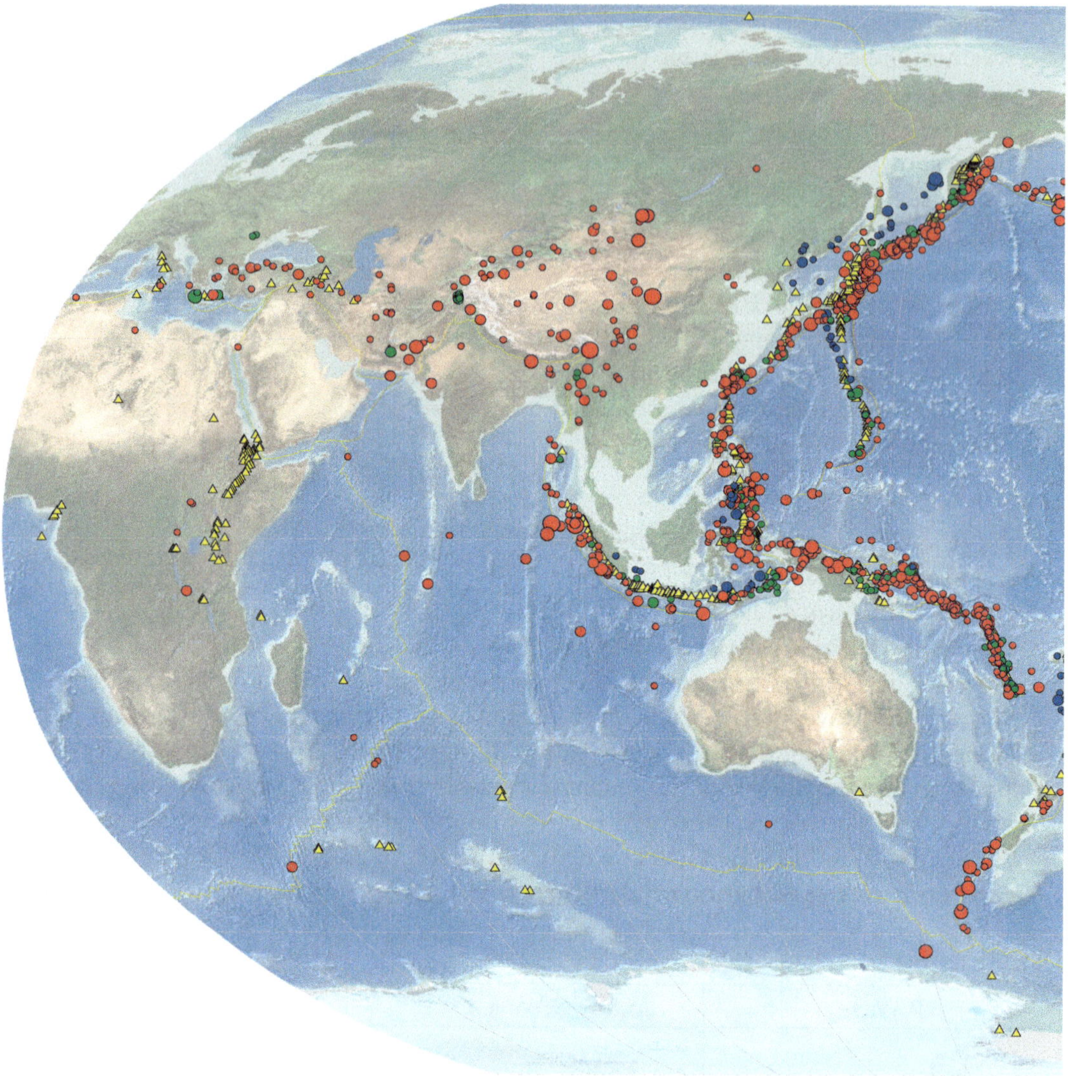

Earthquake Magnitude

- 7.0–7.5
- 7.5–8.0
- 8.0–8.5
- 8.5+

△ Active Volcanoes

7.40 World-wide distribution of major earthquakes for 1900–2013. Note the similarity of this map to the map of Earth's volcanoes in Figure 7.1.

S waves (or secondary waves) are the second fastest type of seismic wave, and cause particles in rocks to move back and forth at a right angle to the direction the wave is travelling. S waves can move through solids, but cannot move through liquids.

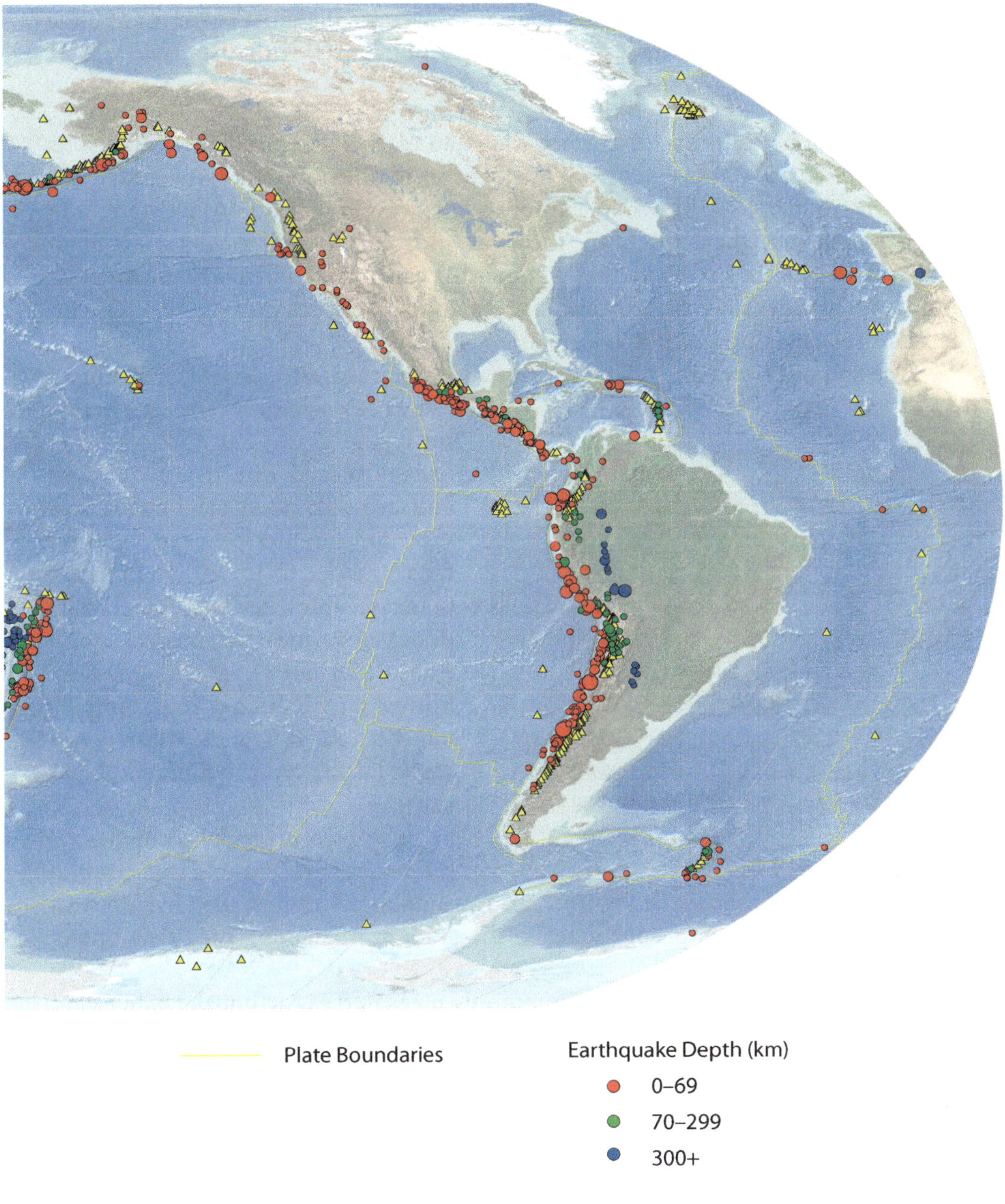

The third type of seismic waves is surface waves. Surface waves travel along Earth's surface, and are similar to waves on bodies of water, so particles move up and down as well as back and forth. During an earthquake, surface waves can make people feel like the ground is rolling. Surface waves are the most powerful of these three wave types, but also have the greatest loss of intensity with distance.

7.4.4. Measuring and Locating Earthquakes

The point beneath Earth's surface where an earthquake originates is called its *focus*. Seismic waves move out from the focus in all directions, like waves on a still pond when a pebble is thrown in, except seismic waves move out from the focus in three dimensions. The depth of the focus ranges from a few kilometers up to about 700 km, well within Earth's mantle. Most strong earthquakes are fairly shallow. The point on Earth's surface directly above the focus is known as the *epicenter*. The focus and epicenter are illustrated in Figure 7.41.

The intensity or strength of earthquakes is measured using an instrument known as a *seismometer*, illustrated in Figure 7.42. Seismometers are usually placed directly on bedrock and are able to detect and record vibrations in Earth's crust that are far too weak to be felt by humans. Some seismometers are specifically designed to measure vertical vibrations; others are designed to measure horizontal vibrations. This is important because P, S, and surface waves shake Earth's crust in different directions.

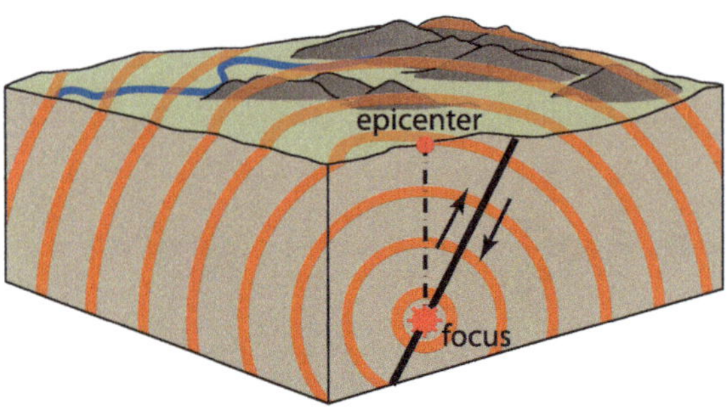

Figure 7.41. The focus is the point where energy is first released along a fault, and may be at considerable depth within Earth's crust. The epicenter is the point on the surface directly above the focus.

The record produced by a seismometer is called a *seismogram*, a sample of which is shown in Figure 7.43. The amplitude (that is, the physical size) of the seismic waves is proportional to the amount of shaking the seismometer experiences.

The intensity of earthquakes is commonly measured as earthquake *magnitude*, a measure of the amount of energy released by an earthquake. A commonly used scale for magnitude is the *Richter scale*. This scale is based on the amplitude of seismic waves as recorded on a seismogram. Earthquakes measuring less than 3 on the Richter scale are too weak to be felt by humans, and the strongest earthquakes ever measured have had magnitudes of 8.9 on the Richter scale.

The Richter scale magnitude of an earthquake is determined by measuring the amplitude of the largest waves on a seismogram. Each number on the Richter scale represents an earthquake with wave amplitudes ten times greater than the previous number. An earthquake with a magnitude of 7 has waves with an amplitude ten times greater than an earthquake with a magnitude of 6. The differences in actual energy released between earthquakes of different magnitudes is even greater than the amplitudes show: each increase of magnitude represents a 32-fold increase in the amount of energy released. An earthquake with a magnitude of 7 releases 32

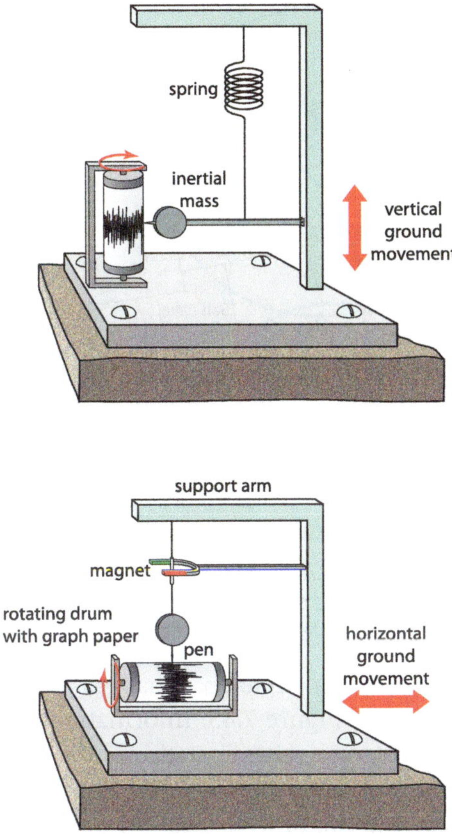

Figure 7.42. Seismometers can be constructed to record horizontal vibrations or vertical vibrations.

Figure 7.43. A seismogram being recorded on paper.

times more energy than an earthquake with a magnitude of 6, and an earthquake with a magnitude of 8 releases 1,024 (= 32 × 32) times more energy than an earthquake with a magnitude of 6. Table 7.2 summarizes the effects of earthquakes of different magnitudes.

Because different kinds of waves travel through Earth's crust at different speeds, P waves, S waves, and surface waves do not arrive at a seismometer at the same time. Geologists determine the distance to the earthquake's epicenter by measuring the time difference between the arrival of the P and S waves. The distance from the seismometer to the epicenter of an earthquake is estimated using a seismic travel time curve, such as the one illustrated in Figure 7.44. At a distance of 1,000 kilometers from an earthquake's epicenter, the gap between the arrival of the first P wave and the first S wave is a little over two minutes. A gap of five minutes, as shown in the figure, means that the epicenter is about 3,800 kilometers away.

Richter Scale Magnitude	Effects Near Epicenter	Number of Earthquakes in Typical Year
<2.9	Not felt, but recorded by seismometer	600,000
3.0 to 3.9	Felt by some	350,000
4.0 to 4.9	Felt by most	6000
5.0 to 5.9	Some structural damage	800
6.0 to 6.9	Significant damage	250
7.0 to 7.9	Extensive damage	20
>8	Catastrophic damage	1–2

Table 7.2. Earthquake magnitudes using the Richter scale.

The distance to an epicenter determined from a seismic travel time curve can

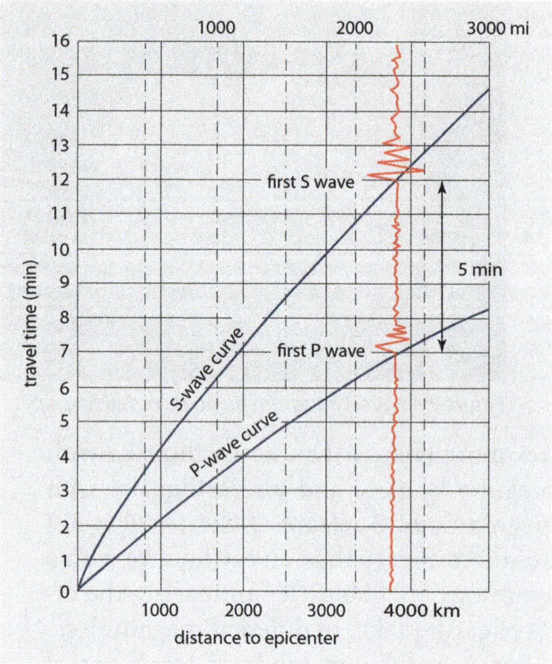

Figure 7.44. The distance from a seismometer to an earthquake's epicenter can be calculated using a seismic travel time curve.

Figure 7.45. An earthquake's epicenter is at the intersection of the three circles that represent distances from three seismic stations.

be plotted as a circle on a map, as illustrated in Figure 7.45. Information from just one seismic station, however, is not sufficient for determining the location of the epicenter. Geophysicists determine the precise location of an earthquake's epicenter by using the travel times recorded at seismometers in three different locations.

Learning Check 7.4

1. What causes earthquakes?
2. What is elastic rebound and how does it relate to earthquakes?
3. Describe the three types of seismic waves.
4. What does the Richter scale measure?
5. What is a seismic travel time curve and how can it be used to determine the distance between a seismometer and an epicenter?
6. What is the difference in arrival time for P waves and S waves at a distance of 1,000 km from the epicenter? What about distances of 2,000 and 3,000?
7. Why is it necessary to have data from at least three seismic stations in order to locate an earthquake's epicenter?

7.5 Earthquake Damage

The strongest earthquakes shake the ground so severely that most buildings are either seriously damaged or destroyed. These earthquakes last for up to three or four minutes. The amount of damage from an earthquake depends on a number of factors, such as the magnitude of the earthquake, the depth of the focus, the nature of the ground—solid rock or shifting sediments—and the quality of construction of buildings. Fortunately, many major earthquakes occur in remote areas and result in few deaths and little damage. Table 7.3 describes the modified *Mercalli intensity scale*. The Richter scale is a measure of earthquake energy, but the Mercalli scale describes an earthquake's damage.

Intensity	Shaking	Description
I	not felt	Not felt except by a very few under especially favorable conditions.
II	weak	Felt only by a few persons at rest, especially on upper floors of buildings.
III	weak	Felt quite noticeably by persons indoors, especially on upper floors of buildings. Many people do not recognize it as an earthquake.
IV	light	Felt indoors by many, outdoors by few during the day. At night, some awakened. Dishes, windows, doors disturbed. Sensation like heavy truck striking building.
V	moderate	Felt by nearly everyone; many awakened. Some dishes, windows broken. Unstable objects overturned.
VI	strong	Felt by all, many frightened. Some heavy furniture moved; a few instances of fallen plaster. Damage slight.
VII	very strong	Damage negligible in buildings of good design and construction; slight to moderate in well-built ordinary structures; considerable damage in poorly built or poorly designed structures.
VIII	severe	Damage slight in specially designed structures; considerable damage in ordinary substantial buildings with partial collapse. Damage great in poorly built structures. Fall of chimneys, factory stacks, columns, monuments, walls.
IX	violent	Damage considerable in specially designed structures. Damage great in substantial buildings, with partial collapse. Buildings shifted off foundations.
X	extreme	Some well-built wooden structures destroyed; most masonry and frame structures destroyed.
XI	extreme	Few, if any, masonry structures remain standing. Bridges destroyed. Broad fissures in ground.
XII	extreme	Damage total. Waves seen on ground surfaces. Objects thrown upward into the air.

Table 7.3. The modified Mercalli intensity scale.

7.5.1. Earthquake Intensity

Sometimes there is only a general correlation between earthquake magnitude, as measured by the Richter scale, and earthquake effects, as measured by the Mercalli scale. For example, an earthquake near Seattle and Tacoma, Washington in 2001 measured 6.8 on the Richter scale, but caused mostly minor damage, and didn't result in any loss of human life. On the other hand, an earthquake with a magnitude of 6.6 hit the city of Bam, Iran in 2003, killing over 26,000 people. The difference between these two earthquakes was building construction. Major buildings in the Seattle area have to meet strict earthquake building codes, and most houses are built out of wood. Wood is flexible and withstands a considerable amount of shaking. Steel is also flexible, so steel structures usually fare well in earthquakes. Most structures in Bam, shown in Figure 7.46, were made out of brick without any kind of reinforcement that would keep the bricks together when shaken severely. When the earthquake hit in the middle of the night, the buildings collapsed, and most victims were crushed under the weight of their homes.

Figure 7.46. Bam, Iran, before and after the 2003 earthquake.

Another factor that affects the amount of earthquake damage is the type of foundations on which structures are built. Structures built on solid bedrock tend to withstand shaking much better than structures with foundations in loose soil. The worst situation occurs when buildings are on sediments that are saturated with water. When earthquake waves hit water-saturated sediments, the sediments behave as a liquid. Figure 7.47 shows apartment buildings that toppled during an earthquake in Japan in 1964 due to water-saturated soil that became liquid-like and could not support the buildings.

Figure 7.47. These buildings in Japan stayed intact as they toppled during an earthquake in 1964. The buildings were built on unconsolidated deposits that had become saturated with water.

In addition to shaking of the ground, other hazards associated with earthquakes

Figure 7.48. Much of the damage done to San Francisco in the 1906 earthquake was due to fires that burned out of control for several days.

include landslides, flooding, and fires. The greatest amount of damage in San Francisco during the great 1906 earthquake was due to fires that started during the earthquake. The fires quickly burned out of control, as shown in Figure 7.48. Because underground water lines were broken during the earthquake, firefighters had no water to put out the flames and the fires burned for several days.

7.5.2. Tsunamis

One of the most devastating phenomena associated with earthquakes are *tsunamis*.[8] A tsunami is a large ocean wave or series of waves caused by events such as earthquakes, volcanic eruptions, and landslides. These events must cause some sort of vertical motion of the seafloor—pushing large quantities of water up or down—in order to cause a tsunami. In the open ocean, tsunami waves can travel at speeds up to 800 kilometers per hour, but have amplitudes of about one meter, so a person on a ship would not even notice the tsunami passing. However, when tsunamis enter shallow water along coastlines, the waves slow down considerably while growing greatly in amplitude at the same time. This results in wave heights sometimes exceeding 30 meters (about 100 feet).

Sometimes when a tsunami strikes a shoreline, one of the first things people notice is actually the opposite of what one might expect: a dramatic withdrawal

8 *Tsunami* can pronounced with or without the "t."

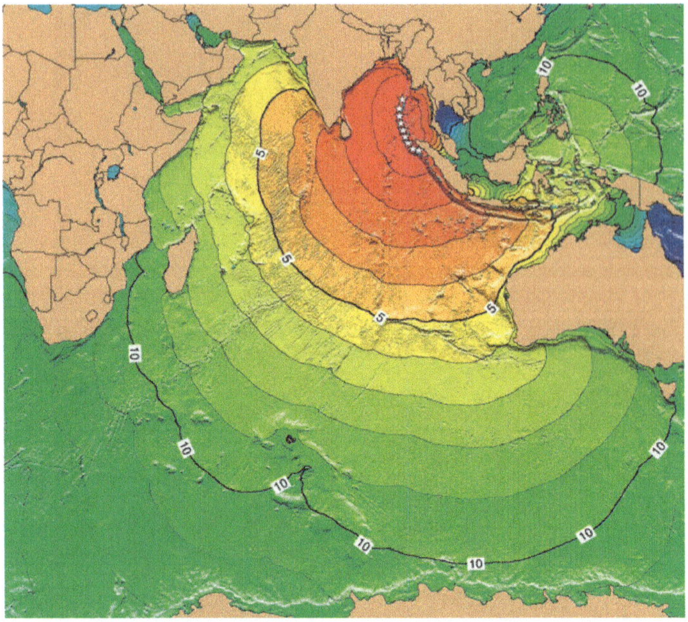

Figure 7.49. Map showing travel times for the 2004 Indian Ocean tsunami. The colors do not indicate wave height, but the number of hours the tsunami travelled. The fault, marked with stars, slipped along a deep sea trench.

of water from the coast, exposing land that is normally underwater. Soon afterward, however, the water level begins rising quickly and then seawater surges inland, flooding far inland if the coastline is flat. The incoming water is typically not a wall of water, but a relentless rising surge. Tsunamis usually come in multiple waves separated by an interval of 10 to 15 minutes.

Two especially devastating tsunamis have occurred so far in the 21st century. In 2004, a powerful earthquake in the Indian Ocean near Indonesia

Figure 7.50. The 2011 tsunami in northern Japan destroyed many coastal communities and carried large ships several kilometers inland.

unleashed a tsunami that killed an estimated 230,000 people—the deadliest tsunami in history. Figure 7.49 is a map showing how the tsunami spread throughout the Indian Ocean, killing people as far away as South Africa. In 2011, another powerful earthquake struck northern Japan, creating a tsunami that had wave heights over 40 meters (130 feet) in places, and killed over 18,000 people. Figure 7.50 shows damage from the Japan tsunami.

The most severe damage from the Indian Ocean and Japanese tsunamis was close to the epicenters. But tsunamis are able to travel across vast stretches of ocean. A strong earthquake in Chile, in South America, in 1960, produced a tsunami that killed 200 people in Japan 23 hours later.

Learning Check 7.5

1. Compare the Mercalli intensity scale to the Richter scale.
2. Give some reasons why an earthquake with a moderate magnitude in one place might cause more damage than an earthquake with a moderate magnitude in another place.
3. Contrast the characteristics of a tsunami in the deep ocean with its characteristics when it comes ashore.

7.6 The Interior of the Earth

We have encountered the different parts of the interior of Earth—core, mantle, and crust—several times in this book. Now we have the background we need to take a closer look at the structure of the inside of our planet. Figure 7.51 diagrams the structure of Earth's interior.

7.6.1. Crust

The *crust* is the outermost layer of the solid Earth, and the only part of Earth that we have direct access to. The crust is exceptionally thin compared to the rest of the planet, comparable (in proportion) to the thickness of the peel on an apple. Continental crust and oceanic crust are quite different from each other. Continental crust is made primarily of rocks with a composition similar to granite, such as granite itself and granitic gneiss, a metamorphic rock. Continental crust is very complex, and includes a significant amount of other igneous, metamorphic, and sedimentary rocks.

Parts of the continental crust are very old; the oldest rocks are about 3.8 billion years old. The deeper granitic and gneissic parts of the crust are referred to as *basement rocks*. In some places these basement rocks are exposed at Earth's surface, but in most places the basement rocks are covered by layers of sedimentary rocks.

Continental crust ranges in thickness from 30 to 70 kilometers. The thinnest parts of the continental crust are under areas of plains, and the thickest parts are beneath major mountain ranges. Mountain ranges actually have roots. The higher

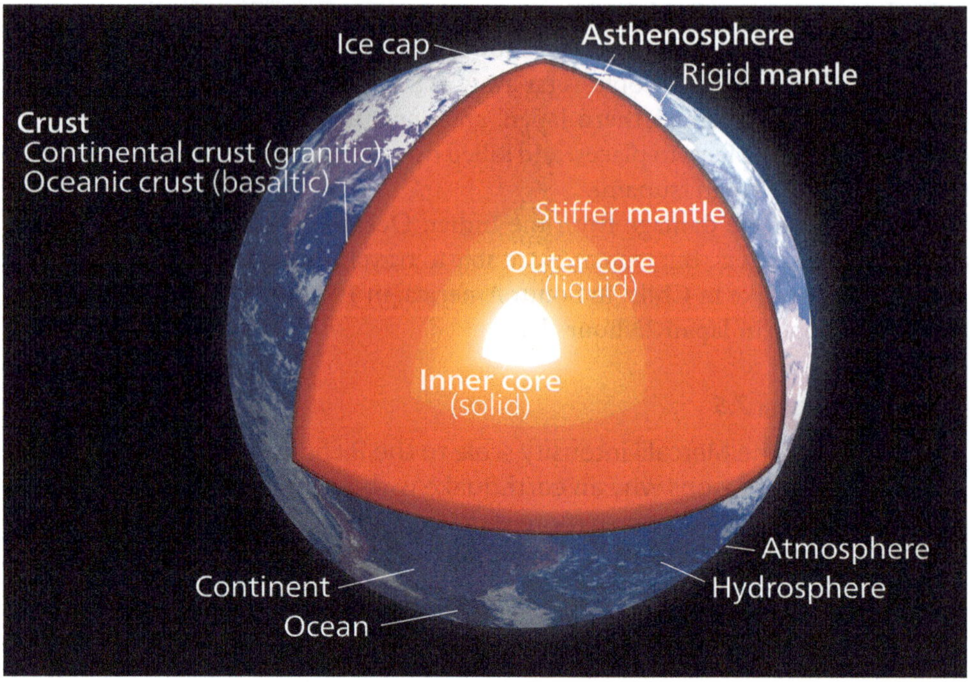

Figure 7.51. Earth's interior, showing the crust, mantle, and core. (This diagram is drawn to scale.)

the mountain range, the deeper the continental crust depresses into Earth's underlying mantle.

Ocean basins are underlain by oceanic crust, which is composed mostly of basalt, and the coarse-grained equivalent of basalt, which is gabbro. Basalt is found in the upper part of oceanic crust and gabbro at greater depth. Oceanic crust is much thinner than continental crust, with a typical thickness of about 10 kilometers.

7.6.2. Mantle

In 1909, a Croatian geophysicist named Andrija Mohorovičić discovered that seismic waves suddenly increase in velocity at a depth of greater than 30 kilometers beneath Earth's surface. The explanation for Mohorovičić's discovery is that there is a distinct change of rock types from granite (under the continents) to a denser rock called *peridotite*, and seismic waves travel faster through denser rock. This break in rock types is now called the Mohorovičić discontinuity, and marks the boundary between Earth's crust and Earth's mantle. The crust-mantle boundary is almost always called the *Moho* for the sake of simplicity.

Earth's *mantle* is the thick layer between the crust and the core. It is composed of dense iron-rich and magnesium-rich silicate rocks such as peridotite. The mantle is almost entirely solid, though some upper portions are partially molten. The molten areas provide magma that rises into Earth's crust.

There are three primary ways in which we can investigate the mantle. The first of these is through seismic waves. By studying how seismic waves bend as they travel through the mantle, geophysicists deduce details about the density and structure of the mantle. The second means is through laboratory studies. For example, scientists subject minerals such as olivine to the extreme pressures that are present in the mantle—pressures ranging from 10,000 to over one million times as great as what is experienced on Earth's surface—and determine how the arrangement of atoms in the crystal structure changes as pressure increases.

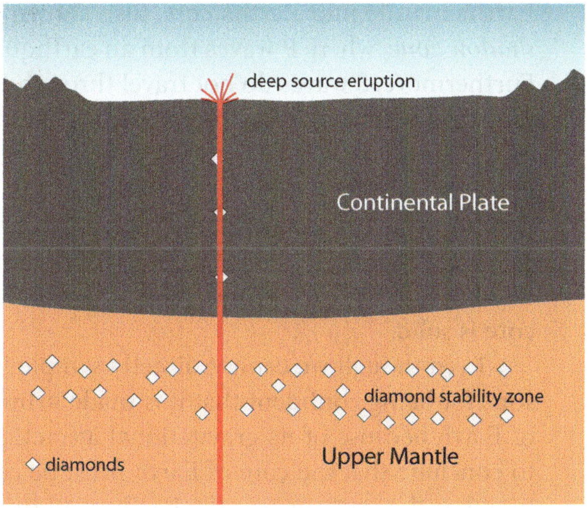

Figure 7.52. Diamonds are lifted from the mantle to the crust through narrow volcanic vents called kimberlite pipes. The rocks in these pipes give us a picture of what kinds of rocks are found in Earth's mantle.

The third way is by examining pieces of the mantle that are available to us at or near Earth's surface. Diamonds can only form under extremely high pressures, such as those found at depths close to 200 kilometers—well within the mantle. Diamonds are brought into the upper crust in narrow volcanic vents known as "kimberlite" pipes, illustrated in Figure 7.52. The material in kimberlite pipes gives us a glimpse at mantle material. It is thought that pyroclastic material was ejected from these volcanoes at supersonic speeds. There are no kimberlite pipe volcanoes active on Earth today, but the eruption of one of these volcanoes must have been spectacular.

Interestingly, deformation of the crust along convergent plate boundaries has sometimes brought slivers of mantle rock up to the surface. Portions of the island of Cyprus are an example.

7.6.3. Core

The velocity of seismic waves increases with pressure, and this causes the paths of seismic waves to curve as they go through Earth's interior, as illustrated in Figure 7.53. Using this knowledge, geophysicists discovered that there is another sharp change in composition within Earth far beneath the Moho. The paths of seismic waves bend sharply[9] when they go from

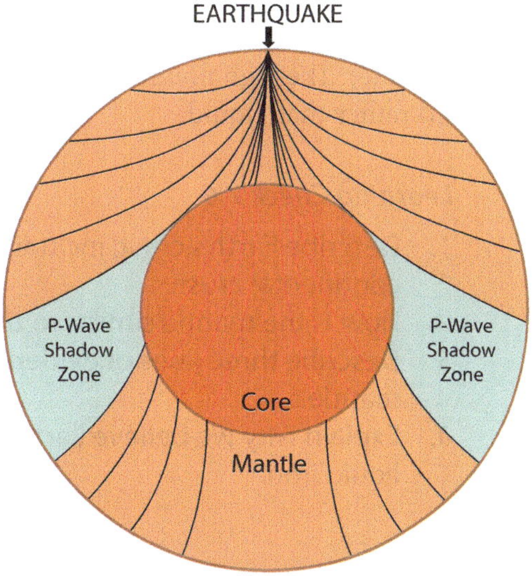

Figure 7.53. The paths of seismic waves curve as they travel through Earth's interior, and bend sharply as they move from the mantle to the core.

9 This bending is called *refraction*.

Earth's mantle into Earth's core, also shown in Figure 7.53. This creates a P wave *shadow zone*, where P waves from an earthquake are not detected by seismometers. Furthermore, S waves do not travel through Earth's core at all, which led geophysicists to conclude that Earth's core is liquid, rather than made out of solid rock.

We now know that Earth's core is divided into an outer core and an inner core. Both parts of the core are composed largely of iron, probably with a significant amount of nickel and perhaps oxygen and sulfur. The temperature of the core ranges from about 4,000°C at the core-mantle boundary, up to over 5,400°C at greater depths—hotter than the surface of the sun! The outer core is liquid, and the inner core is solid.

Even though we cannot directly sample Earth's core, we have a number of reasons for being confident that it is made primarily of iron. First, we know the mass of Earth because of its gravitational attraction for other objects, and this leads us to conclude that the core of Earth must be considerably denser than the materials in either the crust or mantle. Earth would not have as much mass if its core were composed out of silicate rocks such as granite, gabbro, or peridotite. Second, we know that Earth has a magnetic field and that it is generated deep inside the planet. This is explained by a liquid iron core in the Earth, with magnetic properties. No other metal can explain Earth's magnetism. Third, we know that some meteorites are made out of iron and nickel, and many others are rocky with a composition similar to Earth's mantle. If, as scientists suppose, Earth formed from a collection of meteor-like material, it makes sense to conclude that Earth's inner composition is similar to that of meteorites.

The inner core is difficult to study, but we know that P waves travel through the inner core at speeds compatible with what a solid core would produce. Its temperature is about the same as that of the deepest parts of the outer core, but pressures are greater. The high pressure may be the best explanation for why the inner core is solid rather than liquid.

Learning Check 7.6

1. Describe Earth's crust, including the differences between oceanic and continental crust.
2. How is the mantle similar to the crust, and how is it different?
3. Describe three lines of evidence that allow us to know about Earth's mantle.
4. Explain why we believe Earth's outer core is composed primarily of hot, liquid iron.

Chapter 7 Exercises

Answer each of the questions below as completely as you can. Write your responses in complete sentences unless instructed otherwise.

1. In what ways might volcanoes be considered to be part of God's good creation?
2. Describe the three main types of volcanoes in terms of their size, shape, rock types, and style of eruption (explosive or non-explosive).
3. Two earthquakes have magnitudes of 5 and 8 on the Richter scale. How much larger on a seismograph is the amplitude of the 8 earthquake than the 5 earthquake? How much more energy was produced by the 8 earthquake than the 5 earthquake?
4. Use Figure 7.51 to estimate the approximate volumes of Earth's crust, mantle, and core. What percentage of Earth's total volume is in each of these three regions? What assumptions did you have to make as you did these calculations? (Note: the formula for the volume of a sphere is $V = 4\pi r^3/3$, where r is the radius of the sphere. The radius of Earth is 6,317 km.)
5. Use Figure 7.44 (seismic travel time graph) to answer the following questions:
   a. What is the distance from a seismometer to the epicenter of an earthquake if the difference in arrival times between the P and S waves is 1 minute?
   b. What is the distance from a seismometer to the epicenter of an earthquake if the difference in arrival times between the P and S waves is 4 minutes?
   c. If the epicenter is 5,000 kilometers from the seismometer, what is the time difference between the arrival of the P and S waves?
6. Write a report or prepare a presentation about a specific volcano that has had significant eruptions at some time in the past, or has the potential to have a significant eruption in the future. Your report should include information such as the volcano's name and location, relationship to plate tectonics, type of volcano, eruptive history, and hazards to humans.
7. Write a report or prepare a presentation about an important earthquake that occurred in history. Your report should include the name and type of fault involved, magnitude, damage, and human casualties caused by this earthquake.

Experimental Investigation 4: Studying Volcanoes with Topo Maps

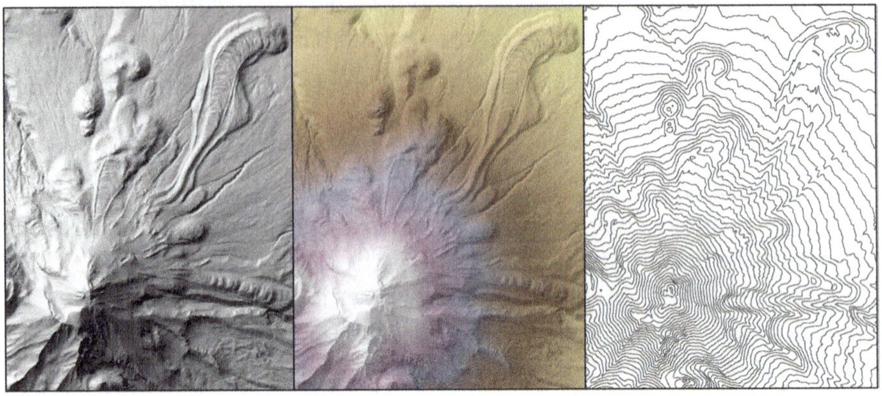

Three computer-generated views of Mt. Shasta, California, which is a composite cone volcano. The first frame shows elevation data with shadows generated as if by the sun in the northwest. The middle frame shows elevation as color bands. The third frame shows contours, with a contour interval of 50 meters.

Overview

The landforms created by volcanoes are clearly expressed on topographic maps. In this activity, you will study shield volcanoes, composite cones, and cinder cones. Pay attention to map scale; the scale required to show an entire shield volcano is quite different from the scale needed to show composite cones or cinder cones.

Basic Materials List

- Three topographic maps: Hawaii, Sitka (A-6), Sunset Crater East
- Ruler
- Graph paper (5 squares to an inch preferred)

Map 1—Hawaii

The largest of the Hawaiian Islands is simply called Hawaii. Hawaii is composed of five shield volcanoes. Kohala is the oldest of the five and is extinct, which means geologists do not expect it to undergo any further eruptions. Mauna Kea is dormant (sleeping) and has not erupted for about 4,600 years, but will likely erupt again sometime in the future. Hualalei last erupted in 1801, and will likely erupt again. Mauna Loa is the largest volcano on Earth. Basalt lava from Mauna Loa is low in silica, so it flows easily for long distances. Mauna Loa last erupted in 1984. Kilauea may be the most active volcano on Earth, and has been erupting continuously since 1983. Many of Kilauea's lava flows reach the Pacific Ocean.

1. What is the scale of this map? What is the contour interval?
2. Find Kohala on the map. What evidence do you see that Kohala is being eroded by streams?
3. Determine the gradients of Mauna Kea and Mauna Loa, measured from the 6,000-foot contour up to the summit of each. Express your results in feet per mile and percent slope. Which volcano has a steeper gradient? How could you have known which volcano was steeper by just looking at the map?

4. Find the lava flow of 1859, northwest of the summit of Mauna Loa. This lava flow traveled all the way from the summit to the Pacific Ocean near Keawaiki. Determine the length, elevation change, and average gradient for the 1859 lava flow.
5. There is considerable variation in how fast lava flows travel across the landscape at Mauna Loa, ranging from 0.1 miles per hour (or even slower) up to 6 miles per hour (about twice the speed of a brisk walk). How long could it take for lava to flow from Mokuaweoweo, which is the summit caldera of Mauna Loa, to the city of Hilo?

Map 2—Sitka (A-6), Alaska

Mt. Edgecumbe is a composite cone in southeastern Alaska. The most recent eruption was about 4,000 years ago.

1. What is the scale of this map? What is the contour interval?
2. Draw a topographic profile of Mt. Edgecumbe, following a line that goes from the northwest through the summit to the southeast.
3. Propose a hypothesis for why the slope changes with distance from Mt. Edgecumbe's summit.
4. Determine the gradient (feet per mile and percent slope) of Mt. Edgecumbe's upper slopes (above 1,500 feet) and lower slopes (below 1,000 feet). Compare these values to those you determined for the shield volcanoes on Map 1, and explain why they are different.

Northwest of Mt. Edgecumbe is a circular feature with contour lines that look different. These contours are called *depression contours* and are symbolized with short spikey lines on them to show that the feature is a hole rather than a hill. The short lines on the contours are called *hachures* and they always point downhill (toward the bottom of the hole). The summit crater of Mt. Edgecumbe also has depression contours.

5. What is the diameter of this feature? How deep is the hole, measured both from the lowest and highest elevation around its rim?
6. Propose a hypothesis for how this feature formed.

Map 3—Sunset Crater East, Arizona

Sunset Crater is a cinder cone in Arizona that formed in an eruption in the late 11th century. The eruption forced Native Americans in the area to temporarily abandon some of their communities, but in the longer term, volcanic ash improved the water-holding capacity of soil for their crops. Sunset Crater is only partially shown on this map, up in the northwest corner. Most of the hills on the map are cinder cones that are older than Sunset Crater.

1. What is the scale of this map? What is the contour interval?
2. What is the gradient (in feet per mile and percent slope) of Stewart Crater, in the northeast part of the map? How does this compare to the gradient values for Mauna Loa and Mt. Edgecumbe?
3. If these cinder cones were mapped at 1:250,000 (the same scale as the Hawaii map), how would they look on the map?

Chapter 8

Weathering, Erosion, and Soils

The Great Plains region of North America stretches from Texas northward to the central provinces of Canada. Much of the Great Plains does not receive sufficient precipitation to support extensive forests, so most of the plains are covered by grasslands. In the late 1800s and early 1900s, large areas of the prairies were converted to dry farmland, meaning that the farmers were dependent on rain and snow to water their crops. In their natural state, the grassland soils were protected from significant wind erosion by a thick network of grass roots. When planted with crops such as wheat, however, the land was plowed every year or every other year, and the soil sat barren for parts of the year. This farming practice worked well in the wetter-than-normal years in the early decades of the early twentieth century, but when intense drought hit in the 1930s, the crops withered, leaving barren fields with no grass roots to hold the soil in place, resulting in massive soil erosion and dust storms. The dust was so thick in the air at times that the storms were called "black blizzards." The hardest hit areas—in parts of Oklahoma, Texas, and Kansas—became known as the "Dust Bowl." The combined hardships of the Dust Bowl and the Great Depression left many farmers penniless.

Wetter conditions returned by the end of the 1930s, and farmers adopted new practices that reduced wind erosion. These practices included planting rows of trees to slow down the wind, reducing the amount of time the ground is bare, and increased pumping of groundwater to irrigate crops. These practices reduced, but did not eliminate, the risk of serious soil erosion.

Objectives

After studying this chapter and completing the exercises, you should be able to do each of the following tasks, using supporting terms and principles as necessary.

1. Explain the differences between mechanical and chemical weathering, and describe the various types of each.
2. Explain why the breakdown of rocks through mechanical weathering makes it easier for chemical weathering to occur.
3. Describe the relationship between weathering, erosion, and deposition.
4. Explain how differential weathering shapes landscapes.
5. Define *soil* and explain why soil is understood to be much more than simply loose mineral grains.
6. Use the soil texture triangle to classify soils according to soil texture.
7. Describe a soil profile with its A, B, C, and O horizons.
8. Explain why soil conservation is important and describe the techniques farmers use to conserve soil.

Vocabulary Terms

You should be able to define or describe each of these terms in a complete sentence or paragraph.

1. Acid
2. Chemical weathering
3. Desertification
4. Differential weathering
5. Dissolution
6. Erosion
7. Frost wedging
8. Gully
9. Humus
10. Hydrolysis
11. Mechanical weathering
12. Oxidation
13. Rill
14. Soil
15. Soil conservation
16. Soil horizon
17. Soil profile
18. Soil texture
19. Spheroidal weathering
20. Talus
21. Topsoil
22. Unloading

8.1. Mechanical and Chemical Weathering

In previous chapters, you learned about the uplifting of mountain ranges through processes such as the folding and faulting of rocks and volcanic eruptions. Even as mountains are being formed, the actions of weathering and erosion work to tear those mountains down. The material eroded from mountains and hills is transported by wind, water, gravity, and glaciers, and is eventually deposited in low places, such as lakes, floodplains, coastal plains, and ultimately in the oceans. Weathering is an ongoing process that is an important part of the rock cycle.

Weathering was defined earlier in this book as the breakdown of rocks to form loose sediments. A good place to observe weathering in progress is human-made substances such as the concrete in sidewalks. Over time, concrete begins to chip, flake, and crack. If sidewalks are not maintained, grass and weeds begin to grow in the cracks. In many climates, even trees grow in the cracks, causing displacement of large pieces of the sidewalk. Given enough time—perhaps decades or centuries—there would be little evidence that the sidewalk ever existed.

There are two main categories of weathering: mechanical and chemical. *Mechanical weathering* involves processes that break rock into smaller pieces without changing the chemical composition of the minerals in the rock. An example of mechanical weathering is the breakdown of granite to sand grains composed of quartz, feldspar, and other minerals. *Chemical weathering* involves the transformation of the original minerals into new minerals, such as the weathering of feldspar to form clay, or the weathering of olivine to form iron oxides.

8.1.1. Mechanical Weathering

There are three primary means by which rocks are broken down by mechanical weathering: frost wedging, unloading, and biological activity. Individual mineral grains are further broken down by mechanical weathering as they are transported by water, wind, or ice.

Figure 8.1. Frost wedging occurs when water within cracks in rocks freezes and expands. This is a purely mechanical weathering process since no chemical changes occur to the minerals in the rock.

Figure 8.2. This conical-shaped boulder deposit is an example of talus. Talus is common in mountainous areas that experience frequent freezing and thawing.

Frost wedging, illustrated in Figure 8.1, is the mechanical weathering of rocks by the formation of ice in cracks. Unlike most substances, water expands when it freezes. Frost wedging occurs when water in cracks in rocks freezes and expands. Frost wedging is a significant factor in many mountainous areas where there is a daily cycle of freezing and thawing for long portions of the year. As rocks at the top of a steep slope break apart due to frost wedging, they often roll down the slope and accumulate in piles called *talus*

Figure 8.3. Half Dome, a prominent mountain in Yosemite National Park in California, is weathering by unloading. Layers in the rock peel away like layers of an onion.

deposits, such as the steep slope of boulders shown in Figure 8.2. (Note, however, that not all talus deposits are the result of frost wedging.)

In areas where temperatures drop to freezing, frost wedging is a primary cause of potholes on roads. When asphalt is new, the tar in the asphalt is flexible, but as the tar ages it becomes hard and brittle and the pavement begins to crack. Water that has soaked into the gravel beneath the road through these cracks freezes when it gets cold, and as temperature fluctuates between temperatures above freezing and below freezing, the pavement is pushed up and down and eventually cracks further, leading to the disintegration of the road surface.

Unloading is the fracturing of rocks due to a reduction of pressure on the rocks when overlying rocks are eroded away. As unloading occurs, some types of rocks fracture in sheets parallel to Earth's surface. When exposed at Earth's surface, these sheets begin to peel away like the layers of an onion. This layered erosion often creates domed surfaces, such as Half Dome in Yosemite National Park, shown in Figure 8.3.

Figure 8.4. Tree roots exert sufficient force to widen existing cracks in rocks.

A third means of mechanical weathering is biological activity. When tree roots grow in cracks in rocks, they exert enough force to widen fractures and eventually split boulders apart, as shown in Figure 8.4. This root wedging is similar to frost wedging except that it occurs in all climates.

A fourth, small-scale type of mechanical weathering occurs to individual grains (particles of all sizes) as they are being transported by water, wind, gravity, or ice. As sand is carried along in a stream, for instance, individual grains rub against each other. The result is that angular edges get broken off, and the grains become more and more rounded as they are transported farther, as illustrated in Figure 8.5. This rounding of grains is especially efficient when grains are transported by wind, or on beaches where grains are continuously rolled back and forth by waves. When studying ancient sandstones, the degree of roundness of the sand grains is one clue geologists use when trying to determine how the sandstone layer formed.

Figure 8.5. The angular sand grains (quartz, rock fragments, shell fragments) at the top have not experienced much mechanical weathering, and have only been transported by water or wind a short distance, but the well-rounded quartz grains in the lower image have been transported a great distance.

8.1.2. Chemical Weathering

Chemical weathering transforms one group of minerals to another—such as the alteration of feldspars and micas to clays—or even completely dissolves minerals. Chemical weathering usually involves water in one way or another.

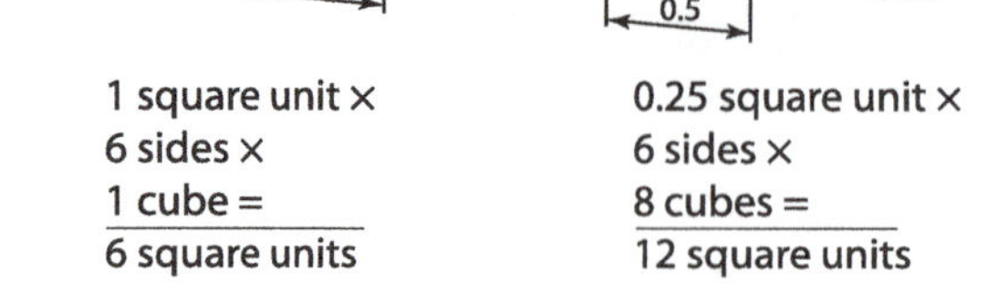

Figure 8.6. Chemical weathering occurs on surfaces exposed to water. When mechanical weathering fractures rock into smaller pieces, the surface area of rock exposed to chemical attack increases.

Water moves easily through most cracks in rocks, but does not seep into the boundaries between individual mineral grains as readily. Because of this, most chemical weathering occurs on the surfaces of rocks or along cracks, rather than deep within rocks. When mechanical weathering breaks rocks into smaller pieces, it makes it far easier for chemical weathering to occur. To illustrate with simplified shapes, in Figure 8.6, the block of rock on the left does not have any cracks, and chemical weathering only works on the outside of the rock. The rock on

Figure 8.7. The most intense chemical weathering often occurs at the intersections of fractures. In these granitic rocks in Joshua Tree National Park in California, fractures have led to spheroidal weathering.

the right has fractures that allow water into the interior of the rock, thus increasing the amount of chemical weathering that occurs. The rate of chemical weathering is directly proportional to the surface area exposed to water. If the rock is a cube with edges one meter long, then the rock on the left has one square meter of surface area per side with six sides for a surface area of 6 m^2. As you can see in Figure 8.6, the total surface area of the eight cubes of the rock on the right is 12 m^2. By adding just three fractures to the rock, the exposed surface area is doubled, so the rate of chemical weathering doubles as well.

When a rock is fractured into blocks as in Figure 8.6, the corners of the fractured rocks are exposed to fluids that might be moving through the rock in any of several different directions. As a result, chemical weathering is most intense at the corners where fractures intersect. This leads to a special type of weathering called *spheroidal weathering*. In this process, initially rectangular blocks become round boulders over time. In Figure 8.7, the rocks higher up are more rounded, while the rocks farther down are still composed of angular blocks.

The simplest form of chemical weathering is *dissolution*—the dissolving of minerals by water. Halite (table salt) easily dissolves in water, breaking it apart into sodium ions (Na^+) and chloride ions (Cl^-). Once dissolved, the ions are carried away by the water, ultimately ending up in the ocean. Because of its high solubility, halite only crops out at the surface in desert areas. Some other minerals are moderately soluble, such as calcite, the main component of limestone. Calcite ($CaCO_3$) slowly dissolves to form calcium ions (Ca^{2+}) and carbonate ions (CO_3^{2-}). These ions move freely in water through soil or into streams. Other minerals, including most silicate minerals, do not dissolve in water to any significant degree.

Hydrolysis is the breakdown of minerals by a chemical reaction with water. While dissolution merely breaks ionic compounds apart into ions, hydrolysis actually changes the minerals. Hydrolysis is the primary means of the chemical weathering of silicate minerals such as olivine, mica, feldspar, and quartz. When a mineral grain is weathered by hydrolysis, some elements are dissolved and carried away in the water and others stay behind to form new minerals, such as a variety of different types of clay.

Oxidation occurs when atoms or ions lose electrons. What that usually means for chemical weathering is that metal ions combine with oxygen to form new minerals. A familiar example of this reaction is when iron oxidizes to form rust. In nature, iron-containing minerals such as olivine and pyroxene are weathered by hydrolysis, which frees up iron ions that oxidize to form oxide minerals such as hematite (Fe_2O_3). Oxidation—whether on an iron nail or a mineral such as olivine—occurs slowly in dry environments but more rapidly in moist settings. This process sometimes causes a reddish coating to form on the surfaces of igneous rocks.

All of these chemical weathering processes occur at faster rates if the water is acidic. There are several definitions of what an *acid* is, but for our purposes, an acid is a water solution that contains an excess of hydrogen (H^+) ions. These hydrogen ions attack and break chemical bonds, accelerating the breakdown of minerals. Water can be naturally acidic for a number of reasons, such as the presence of carbon dioxide in the atmosphere. Atmospheric CO_2 dissolves in rainwater to form carbonic acid. Water can be made more acidic through human activities, such as mining of sulfide ores and burning of fossil fuels, especially coal. Both coal and sulfide ores, such as pyrite (FeS_2) and galena (PbS), contain sulfur, which reacts with water to form sulfuric acid.

8.1.3. Rates of Weathering

Mechanical and chemical weathering work hand-in-hand to break down rocks. As we have seen, when rocks are broken into smaller pieces by mechanical weathering, it is much easier for chemical weathering to act on the rock. Some minerals weather more rapidly than others. In fact, the order of the rates at which silicate minerals weather is the same as the order in which they crystallize from a magma, which was shown on page 113:

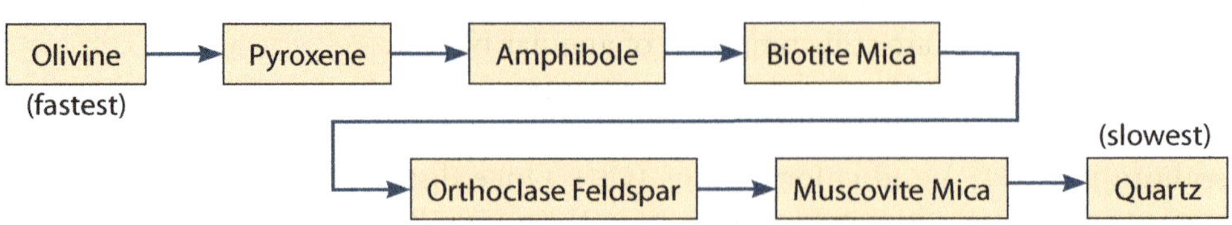

In general, the minerals that form at the highest temperatures, such as olivine and pyroxene, are the most rapidly weathered at the relatively low temperatures that exist on Earth's surface. It is unusual to find olivine, for instance, in sediments or sedimentary rocks, because when olivine weathers out of igneous rocks it is quickly broken down into other substances. At the other end of the scale, quartz is one of the last minerals to crystallize from magma, and is quite stable on Earth's surface. Grains of quartz weathered and eroded out of the Rocky Mountains are transported by streams all the way to the Gulf of Mexico or Pacific Ocean and deposited in deltas, beaches, or offshore. The sand grains might become a little more rounded as they are transported but they are affected very little by chemical weathering.

Learning Check 8.1

1. Define mechanical weathering and chemical weathering.
2. Explain how frost wedging, unloading, and biological activity work to break down rocks into smaller pieces.
3. How do fractures in rocks aid the work of chemical weathering?
4. Compare and contrast dissolution, hydrolysis, and oxidation.
5. Which silicate minerals are most easily weathered, and which are most resistant to weathering? How does this relate to the order in which minerals crystallize from magma?
6. List several variables that affect the rate of weathering.

8.2. Erosion

Erosion and weathering are not the same process. Weathering is a process that occurs in place as mineral grains are loosened through mechanical weathering or altered through chemical weathering. *Erosion*, on the other hand, involves the removal of weathered materials from one place and movement of those materials to another place.

8.2.1. Agents of Erosion

Erosion is an important part of the rock cycle. There are several agents of erosion, the most important of which is water. Even the force of a single large raindrop, such as the drop splashing in soil in Figure 8.8, can dislodge loosened mineral grains and begin the process of erosion. Of course, a heavy downpour or moving water in a stream has an even greater ability to erode material.

Figure 8.8. Soil being splashed by a single raindrop.

When rain falls on Earth's surface, some of it initially seeps into the ground. When the ground becomes soaked, water begins flowing over the surface as sheets of water, carrying mineral grains and dissolved ions along. As the sheets of water flow across the surface, the water naturally concentrates in narrow channels on hill-

Figure 8.9. Rills forming on a gentle slope.

sides. These channels are called *rills*, and they cut deeper into the hillside than the erosion caused by sheet flow. Deeper channels are called *gullies*. Rills and gullies are shown in Figures 8.9 and 8.10.

Water also causes erosion along the coastlines of oceans or lakes. Waves rolling onto either rocky or sandy shorelines exert a tremendous amount of force, and strongly shape the coastal landscape. Tides and ocean currents also cause a great amount of erosion along shorelines. We will examine shoreline erosion in more detail in Chapter 13.

Figure 8.10. This gully near the Dead Sea in Israel formed by water erosion in less than one year.

Another important agent of erosion is wind. Many arid areas have little or no vegetation and loose particles are easily picked up by the wind. Wind erosion is not usually caused by the wind itself, but by the sandblasting effect of particles carried along by the wind. Figure 8.11 shows how wind erosion can create mushroom-shaped pillars in sandy areas. Wind erosion not only occurs in deserts, but anywhere where loose sediments are exposed at Earth's surface, such as near beaches, river floodplains, and plowed agricultural fields.

One more important erosional agent is glacial ice. As we saw in Section 3.2, thick accumulations of ice, such as in polar or high mountain areas, slowly move downhill under the pull of gravity. This glacial ice picks up loose rocks and sand, and these act to grind the surface as the glacier moves across it. Figure 8.12 shows sediment-rich glacial ice.

Figure 8.11. Blowing sand close to the ground surface has scoured the base of this rock down to a narrow neck.

Figure 8.12. Glaciers move downhill carrying particles ranging in size from clay and silt up to boulders. This is the downhill end of a glacier, where the ice is melting.

8.2.2. Transportation and Deposition

Material can be carried tens, hundreds, or even thousands of kilometers by the various agents of erosion. Fine silt and clay picked up by winds over the Sahara Desert is carried by winds all the way across the Atlantic Ocean and deposited in the Caribbean and South America, as illustrated in Figure 8.13. Sediments carried by rivers can cross entire continents.

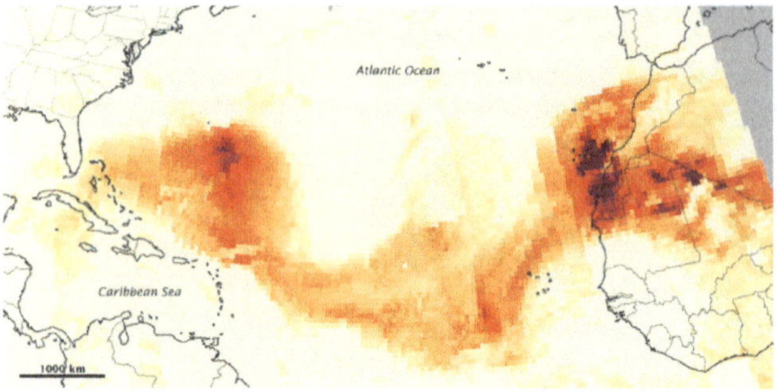

Figure 8.13. A map with data derived from satellite imagery showing dust being blown from the Sahara Desert all the way across the Atlantic Ocean. Darker colors indicate greater concentrations of dust in the atmosphere.

In most cases, transportation of sediments is a step-by-step process. A grain of sand eroded from a hillside in the mountains might be transported down a slope and then sit at the bottom for years. It might then be carried into a stream, where it moves downstream with the current some of the time, but actually spends most of its time trapped in sand or gravel deposits, as shown in Figure 8.14. It might even stay in a deposit for thousands or even millions of years if buried by additional sediments or if the stream changes its course. But the ultimate destination of all eroded sediments is the oceans, where final deposition can be in various environments, such as deltas, beaches, and offshore sand bars. Some sediments are even carried by currents into deep ocean basins. These deposits eventually become rock, and might one day be uplifted through mountain-building processes—all part of the rock cycle.

Figure 8.14. Sediment in streams spends most of its time trapped in deposits such as these gravel bars in Alaska. The gray material throughout the center portion of this image (inside the water boundaries) is gravel.

8.2.3. Differential Weathering and Erosion

Figure 8.15. Gravestones weather at different rates depending on the type of rock and the climate. The letters on this gravestone can no longer be read.

Weathering works at varying rates for different rocks and minerals. This becomes evident by comparing gravestones in an old cemetery, especially in a place with a moist climate. Figure 8.15 shows a gravestone on which weathering has completely removed (*effaced*) the lettering. Granite tombstones usually retain their lettering for centuries, even in moist climates, while marble and limestone tombstones are often unreadable after a century of being exposed to the weather. Chemical weathering occurs more rapidly if rainwater is more acidic (the acid precipitation mentioned previously).

Entire landscapes can be sculpted by *differential weathering*, which results from different rates of weathering on rocks. Coarse-grained felsic rocks such as granite and gneiss are usually resistant to weathering in all climates, and often form

Figure 8.16. Differential weathering in the Grand Canyon, Arizona. Sandstone and limestone are resistant to weathering in dry climates and form cliffs. Shale is more easily weathered in any climate and forms more gentle slopes.

Figure 8.17. The ridges and valleys of the Appalachian Mountains are the result of differential weathering.

elevated features such as hills and mountain ranges. Sandstone layers are usually resistant to erosion as well, while shale layers are usually easily weathered. Limestone is more easily dissolved by water in moist climates, where it forms valleys, while in dry climates limestone is more resistant to erosion and forms ridges. This can be seen in the Grand Canyon in Arizona, shown in Figure 8.16. In the Grand Canyon, layers of sandstone and limestone form the cliffs, but layers of shale form the gentler slopes. The granite and schist at the deepest parts of the Grand Canyon are also very resistant to erosion.

Many hilly and mountainous areas on Earth show the effects of differential weathering. Large parts of the Appalachian Mountains of the eastern United States were formed through the folding of sedimentary rock layers to form anticlines and synclines, which have been deeply eroded over time. The ridges in Figure 8.17 are composed of sandstone or quartzite, and the valleys are formed where shale or limestone have been weathered and eroded.

Many landscapes are more complex than the Grand Canyon or Appalachian Mountains, but differential weathering has been an important process in sculpting much of Earth's surface.

Learning Check 8.2

1. What are the three primary agents of erosion?
2. Distinguish between rills and gullies.
3. Where else does water erosion occur, other than in streams?
4. How does differential weathering work to shape landscapes?
5. Explain the cause of the distinctive shapes in the Grand Canyon.

8.3. Soils

Soil is the loose covering of weathered material that lies on Earth's surface, and is a complex mixture of mineral grains, organic material, water, and air. Soil is one of our most important natural resources, and is a place where the solid Earth, atmosphere, hydrosphere, and biosphere all interact. Soil provides plants with minerals[1] such as calcium, magnesium, and potassium, and other nutrients that plants need for growth. Soil also provides water for plants and a place for roots to hold plants in place.

8.3.1. Soil Composition

A typical soil is mostly composed of mineral matter, air, water, and a smaller amount of organic matter, as illustrated in Figure 8.18. These values can vary considerably from place to place, and even from time to time, as the water content of a soil changes following rainfall or dry spells. For instance, the soils in deserts and swamps, such as those shown in Figure 8.19, are very different from each other: the swamp soil has much higher levels of both moisture and decayed organic matter.

The mineral grains in a soil include a mixture of sand—often quartz grains, but sometimes other minerals or rock fragments—silt, and clay. The percentages of each of these varies considerably from one soil to another, and this affects how well different types of vegetation can grow in the soil. The clay content is especially important, as clay helps the soil to retain nutrients and water. If there is too little clay, the soil cannot retain water and certain nutrients. If there is too much clay, the soil easily becomes compacted, making it difficult for roots to penetrate and organisms such as earthworms to move through the soil.

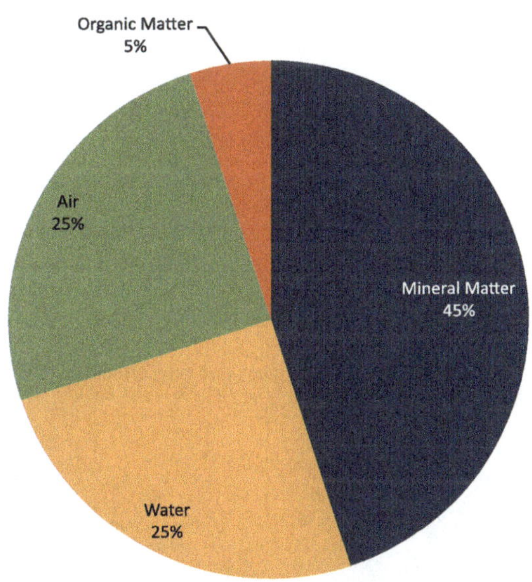

Figure 8.18. Composition of a typical cropland soil.

Recall from Chapter 5 that loose mineral grains can be classified as sand, silt, or clay, depending on their size. *Soil texture* refers to the relative proportions of sand, silt, and clay particles in the soil. A soil scientist will determine the percentage of each of these three size classes and plot them on a triangular diagram called a *soil texture triangle*, shown in Figure 8.20. The soil texture triangle is divided into vari-

1 When speaking about plant nutrition, it is common to include to metal ions such as Ca^{2+}, Mg^{2+}, and K^{+} in the category of minerals.

Figure 8.19. Because the climate is different, the soil in a hot, dry desert (top) is very different from the soil in a cool, moist bog (bottom).

ous regions with names such as loam, sandy clay, and sandy clay loam. For example, a soil with 20% sand, 60% silt, and 20% clay is called a silt loam.[2]

The air and water in soil is in pore spaces between mineral grains. The loose nature of most soils enables plant roots easily to grow down into the soil. Soil organisms, such as microbes and invertebrates, need oxygen and water to survive, and plants that grow in the soil are dependent on soil moisture as well. If the soil becomes compacted, such as when animals walk on it regularly or through human activity, the pore space is greatly reduced.

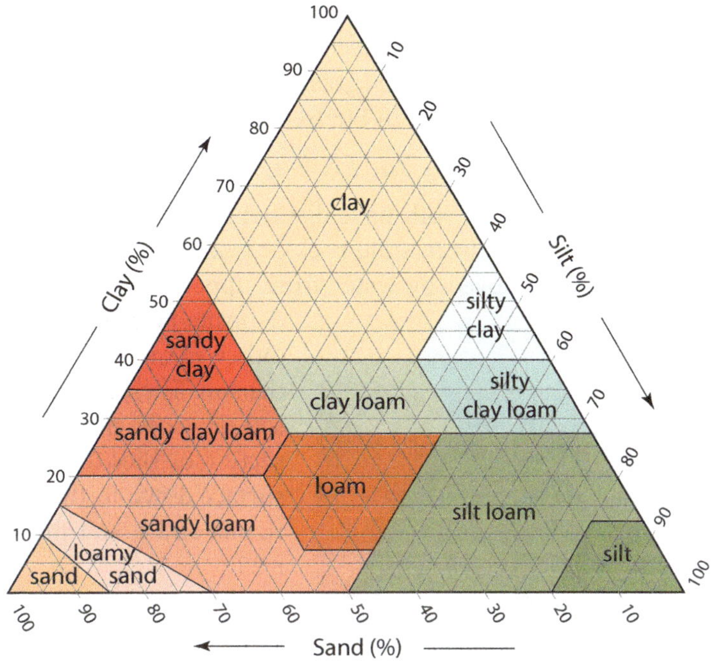

Figure 8.20 The soil texture triangle. Sand percentage (by weight) is plotted using the coordinates along the bottom of the triangle, clay along the left side, and silt along the right side. The soil texture for the soil sample will be where lines for the three values intersect.

Living material in soil includes organisms such as soil bacteria, fungi,[3] mold, various types of worms, a wide variety of insects and other invertebrates; plus the roots of plants. Many of these organisms, such as earthworms and ants, help to keep the soil loosened. Soil bacteria and fungi are important for decomposing the dead remains of organisms, including dead plant debris lying on the ground surface. This decomposition releases nutrients back into the soil that are needed by

Figure 8.21. A gardener's compost is very similar to humus that forms in natural settings as organic material decomposes with the help of soil bacteria and other organisms.

2 The percentages of sand, silt, and clay always add up to 100%.

3 The plural of fungus is either fungi or funguses.

other organisms. Decayed organic material in soil is called *humus*,[4] which typically has a dark brown or black color. Humus is similar to compost that some gardeners create by allowing leaves, grass clippings, and kitchen waste to decompose in a pile or bin, such as shown in Figure 8.21.

8.3.2. Soil Horizons and Profiles

Most mature soils have distinct horizontal zones or layers called *soil horizons*. As a soil forms from solid rock or from loose sediments, some elements and minerals are removed by water from upper levels of the soil and carried down and deposited deeper in the soil, creating this layering. Other elements may be completely removed from the soil. These horizons become more distinct as a soil develops over time. A *soil profile* is a vertical section or slice through soil horizons, from Earth's surface down to the soil's parent material. Soil horizons and a soil profile are illustrated in Figure 8.22.

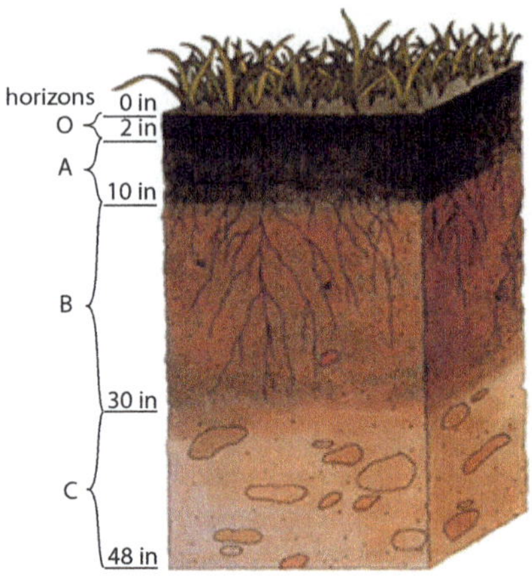

Figure 8.22. A soil profile showing four distinct soil horizons with typical depths.

Most soils have at least three distinct soil horizons, designated from the surface downward as the A, B, and C horizons. The A horizon is composed of mineral matter mixed with humus. The humus in the A horizon gives it its dark color. The A horizon is also characterized by the removal of certain substances, such as iron, clay, and carbonate ions, which are then either deposited deeper down in the soil, or removed entirely. The B horizon is a zone of accumulation of materials removed from the A horizon, so it is often enriched in iron, clay, or carbonate ions. The B horizon is usually noticeably lighter in color than the A horizon. In some B horizons, the accumulation of clay results in a compact layer called a hardpan that does not allow the movement of water down to deeper layers in the soil. In soils in arid regions, the B horizon can be the site of an accumulation of a hard layer of calcium carbonate. The C horizon is a zone of parent material that has been only partially altered by weathering, and is often rocky if the parent material is solid rock.

Some soils have other horizons. A common additional horizon is the O horizon, above the A horizon and composed of loose, partially decayed organic material such as leaves or grass from previous years' growing seasons. Together, the A and O horizons are referred to as *topsoil*. The topsoil is where most of the nutrients needed for plant growth reside, and if the topsoil is lost because of erosion, the soil will be considerably less fertile. The B horizon is sometimes referred to as the subsoil.

4 Pronounced HYOO-mus.

Figure 8.23. As soil develops on bare rock, the vegetation that grows changes: 1) bare rock, 2) lichen and moss, 3) grasses, 4) small shrubs, 5) fast-growing trees, 6) mature trees.

8.3.3. Formation of Soils

Soil forms over time through a combination of weathering and biological activity. There are several factors that work together to determine what kind of soil forms in any given place and how long it takes soil to develop.

One important factor is the parent material, the starting material for creating a soil. The parent material might be solid rock, such as granite, sandstone, or limestone. Granite contains an abundance of both quartz and feldspar, and the weathering of granite results in the production of large quantities of sand, silt, and clay. Limestone, on the other hand, typically contains little quartz, and limestone soils tend to be sand-poor. However, weathering of thick layers of limestone can concentrate what little quartz is there, resulting in a sandy soil. The parent material can also be unconsolidated sediments, such as on a river floodplain. Soil can form more rapidly from sediments than from solid rock, as there is no need for a lengthy period of mechanical weathering.

Another important factor which determines the type of soil that forms, and how rapidly it forms, is climate. As we will discuss in Chapter 16, climate is a function of both temperature and precipitation. Chemical reactions occur more rapidly with higher temperatures. Most of the chemical reactions involved in the breakdown of minerals also occur faster in the presence of water. Soil formation, therefore, occurs most rapidly and thoroughly in hot, moist climates, such as in tropical rainforests, and most slowly in cold, dry regions, such as rocky polar areas. Even though deserts can be hot, the development of soil is impeded in arid areas by the lack of moisture.

Figure 8.24. Lichen on the surface of a boulder.

Figure 8.25. Aerial photograph showing lava flows of various ages on the southwest flank of Mauna Loa, a volcano in Hawaii. The dark areas are lava flows that are less than 200 years old. The lighter areas are weathered lava flows with varying amounts of vegetation.

The intensity of soil formation processes in temperate[5] forests and grasslands, such as in much of the eastern and central parts of the United States, tends to be somewhere in between these extremes.

A third factor that influences the type of soil is time. Figure 8.23 shows the development of a forest over a period of perhaps hundreds of years starting from bare rock. The bare rock might be a fresh lava flow, or it might be exposed to the surface through erosion or a landslide. At first there is no soil, but over time, the interaction of climate and organisms leads to the development of a mature soil. The first organisms to live on the surface of bare rock include lichens, shown in Figure 8.24, and perhaps moss. Lichens can live directly on the rock surface and moss can grow in

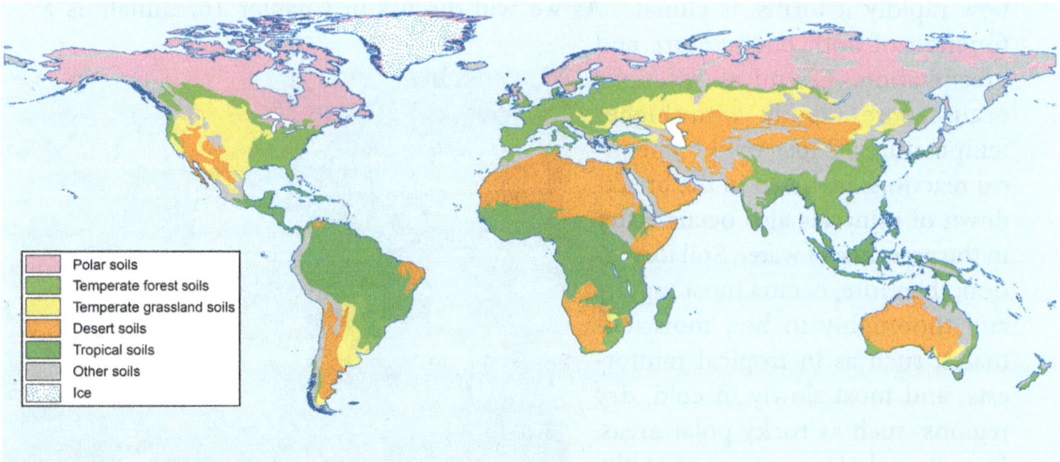

Figure 8.26. A general map of the distribution of major soil types on Earth.

5 *Temperate* refers to climates between the extremes of tropical and polar.

places where wind-blown dust accumulates in cracks and hollows. Lichens produce acid, speeding the chemical weathering of minerals in the rock and producing small amounts of clay and other minerals. After a period of time, additional types of plants, such as grasses, grow in certain places on the thin, patchy soil on the lava flow. Eventually, there will be sufficient soil for the growth of small bushes and then trees, if the climate is sufficiently moist. This process of going from bare rock to mature soil begins immediately in most climates, but takes many decades, centuries, or even thousands of years to run to completion, depending on the parent material and climate. Figure 8.25 shows an aerial image of lava flows in Hawaii. The most recent flows have no vegetation, while other flows have varying amounts of vegetation depending on their age.

8.3.4. Soil Types

Soil scientists have various ways of classifying soils. We have already seen one classification system, which was classification by soil texture, described by the soil

Figure 8.27. A typical polar soil, with a very thin topsoil and frozen soil at depth.

Figure 8.28. A temperate forest soil with B horizon accumulation of clay and iron oxides.

Figure 8.29. Temperate grassland soils have a dark A horizon which is rich in organic matter. These soils usually make excellent cropland.

Figure 8.30. This desert soil has little organic matter, so it is light in color. Soil forming processes occur slowly, so there are still pieces of partially-weathered bedrock at a shallow depth.

texture triangle. Since climate is such a significant factor affecting the nature of soil, another way of classifying soils is based on climate. In this classification, there are five types of soil: polar, temperate forest, temperate grassland, desert, and tropical. Figure 8.26 is a world map showing the distribution of these five soil types. Figures 8.27 through 8.31 show examples of typical soil profiles for these five soil types. Each of the five photos shows the vertical face of a trench cut through the soil.

Polar soils, as shown in Figure 8.27, form at high latitudes such as in land areas surrounding the Arctic Ocean. Polar soils also form in tundra areas at high elevations in mountain ranges. Polar soils are generally thin—only a few centimeters deep—and have weak development of soil horizons. Polar soils are generally underlain by permanently frozen material called permafrost, as shown in Figure 8.32.

Figure 8.31. Tropical soils, whether under forests or savannahs, are typically red or yellow due to high levels of iron oxides.

Temperate climate zones are found at mid-latitude areas such as much of North America, Europe, eastern China, and the southern parts of South America. Temperate forest soils, such as those found in many parts of the eastern United States, are not as fertile as grassland soils. The B horizons of temperate forest soils contain accumulations of clays and iron oxides, making these soils light-colored or reddish, respectively. A temperate forest soil is shown in Figure 8.28.

Figure 8.29 shows a temperate grassland soil. These soils have an abundance of humus in the topsoil, and are exceptionally fertile. Many of the world's most productive agricultural areas, such as the American Midwest, have temperate grassland soils.

Desert soils, such as shown in the soil profile in Figure 8.30, are generally light-colored. They typically have thin A horizons with little organic matter, and B horizons that have accumulations of salts, gypsum, or calcite. Desert soils can be rich in minerals needed by plants, enabling deserts to exhibit a brief burst of plant growth following periods of precipitation.

Tropical soils develop in areas near the equator, either in tropical forests or grasslands. Most tropical areas experience year-round high temperatures and periods of significant rainfall, and tropical soils are the most intensely weathered soils on Earth. Because of this, tropical soils are generally thick and highly altered. Despite the lush plant growth in tropical forests, the soil is typically quite poor in organic matter and nutrients. Heavy rainfall removes most minerals from the soil, leaving behind only those minerals that cannot be dissolved in water, such as aluminum and iron oxides. These oxides give tropical soils a strong yellow or red color, as shown in Figure 8.31. Most of the organic material and nutrients in a tropical forest are in the living organisms rather than in the soil. When plants die, they decompose rapidly and their nutrients are quickly taken up by other plants. If trees are removed from many tropical forests, the soil is exposed to the sunlight and the clay bakes and becomes hard as a brick, making it difficult for plants to grow.

8.3.5. Soil Conservation

Soil erosion is one of the greatest threats to global food production. Loss of topsoil is not always obvious, but is revealed by measurements of topsoil depth and quality over a period of years. The workers in Figure 8.32 are digging into frozen polar soil to measure soil depth. Soil erosion is a natural process, but many natural soils are in a state of equilibrium in which topsoil lost through erosion is replaced by new topsoil through the processes of soil formation. Poor agricultural practices and other human activities have disrupted this equilibrium in many places, meaning that the rate of erosion is now greater than the rate of soil formation.

Figure 8.32. These workers are using a jackhammer to dig into frozen polar soil.

Soil conservation is the prevention of soil erosion or other degradations of soil by adopting techniques such as contour plowing, planting windbreaks, crop rotation, and no-till agriculture. The photo Figure 8.33 shows contour planting, where the farmer plants crops in curved rows that follow contours rather than in straight rows that might go straight up and down slopes. Contour plowing reduces soil erosion by forcing water to flow slowly sideways along a hillside rather than in rills that run straight down the slope. Figure 8.34 shows windbreaks—rows of trees planted to help reduce wind erosion.

Figure 8.33. These crop rows follow contours, which helps prevent rill erosion caused by water running straight downhill.

Figure 8.35 illustrates two soil conservation practices: no-till agriculture and crop rotation. No-till agriculture is a farming practice in which the field is not plowed between harvests. The dead plant debris helps to prevent water loss and soil erosion. Crop rotation is the planting of different crops in alternating growing seasons. In Figure 8.35, there are dead corn stalks on the ground, but the current crop is soybeans. The soybeans help the corn by enriching the soil with nitrogen, which corn is unable to do. The dead corn stalks help the soybeans by holding moisture in the soil.

A variation of crop rotation is to grow multiple crops in the same field. Native Americans throughout much of North and Central America grew what is known as the "three sisters," corn, squash, and beans. These grew together in small mounds of soil. The beans provided nitrogen for the soil, the squash helped to prevent pests and weeds, and the corn provided a pole for the beans to climb. Not only was this combination nutritious for the Indians, but the combination of crops seems to have been sustainable for the soil over periods of many centuries.

Figure 8.34. Windbreaks are rows of trees that help to reduce soil erosion by reducing wind velocity near Earth's surface.

In some parts of the world, soil degradation has led to a process called *desertification*, in which once-productive lands become increasingly desert-like through drought or improper use. Figure 8.36 is a map showing areas that are at risk for desertification. Most of these areas are once-fertile areas on the edges of deserts. The area along the southern edge of the Sahara Desert in Africa is known as the Sahel, and droughts and desertification in this region have led to famines that have affected millions of people. During a period of unusually wet years in the 1960s, governments encouraged people to farm and tend livestock farther north in the Sahel. When the climate became dryer less than

Figure 8.35. Crop rotation in the American Midwest. Corn stalks from the previous year are lying on the ground, helping to reduce erosion and water evaporation. The current crop is soybeans.

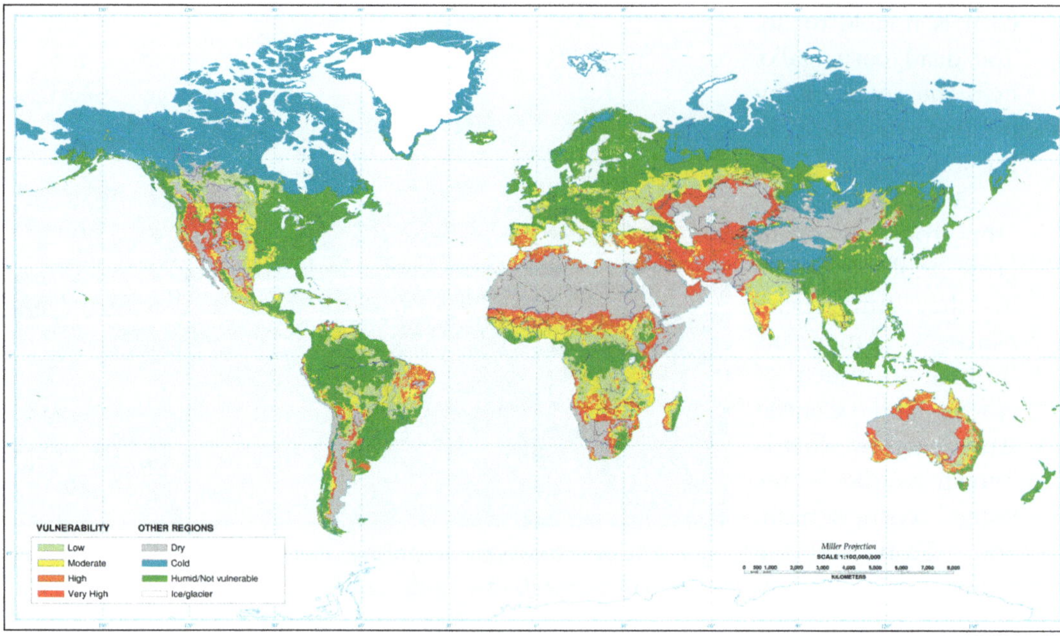

Figure 8.36. Red and orange colors indicate areas at risk for desertification. This does not mean that desertification is occurring at all of these places.

ten years later, these unwise agricultural practices led to desertification, and this resulted in the starvation of many people. This situation was similar to the Dust Bowl in the United States in the 1930s, except that the human consequences were much greater in the Sahel, with widespread starvation and malnourishment. Additionally, the farmers in the Dust Bowl area more successfully adopted new techniques to prevent desertification.

Learning Check 8.3

1. Describe a typical soil in terms of its composition and living components.
2. What is humus?
3. Describe soil horizons and explain how they form.
4. Compare and contrast desert soils, temperate forest soils, and temperate grassland soils.
5. Why is human-caused erosion a problem?
6. Why is soil conservation important?
7. Describe four methods of soil conservation.

Chapter 8 Exercises

Answer each of the questions below as completely as you can. Write your responses in complete sentences unless instructed otherwise.

1. Distinguish between mechanical and chemical weathering.
2. Relate weathering, erosion, and soil formation to the rock cycle.
3. Figure 8.6 illustrated the effect of cutting a one-meter cube of rock into eight smaller cubes, each having sides 50 centimeters long. Calculate how much surface area would be exposed to chemical weathering if the one-meter cube of rock were broken down into tiny cubes, each having edges only one millimeter long.
4. Describe the factors that affect rates of weathering.
5. Investigate differential weathering in the area you live. What types of rocks make ridges and hills? What types of rocks are in the valleys?
6. Describe the various processes involved in separating mineral grains from the side of a mountain and transporting them to a deep ocean basin.
7. Fill in the blank values and use the soil texture triangle in Figure 8.20 to determine the texture of the following samples:

Soil Sample	Percent Sand	Percent Silt	Percent Clay	Texture
1	60	30	10	
2	40	30		
3	15		20	
4		16	23	

8. Make a soil composition pie chart for a soil that is 63% mineral matter, 2% organic matter, 5% water, and 30% air. Which of the five soil types described in Section 8.3.4 is this soil likely to be?
9. Explain the process of soil formation, beginning with bare rock.
10. Speculate on how a soil on the north slope of a hill might be different from a soil on a south-facing slope.
11. Relate soil conservation to the principles of stewardship and sustainability discussed in Chapter 3.
12. Explain the "three sisters" agricultural process used by Native Americans, and describe the benefits provided by each of the types of plants.
13. Why is crop rotation important?
14. Describe how human activities lead to an increase in erosion.
15. Explain the process of desertification.

Chapter 9

Surface Water and Groundwater

Much of the southwestern United States is a desert, but water taken from the Colorado River has transformed parts of this desert into an oasis. Each year, billions of cubic meters of water are diverted from the Colorado River to irrigate cropland, water golf courses, fill swimming pools, and provide drinking water for tens of millions of people in cities such as Los Angeles, San Diego, Phoenix, and Las Vegas. This has been accomplished by building huge dams with reservoirs that store vast amounts of water, as well as a number of canals to carry water to distant cities.

Starting in the 1920s, as both population growth and development of agriculture was accelerating across the region, water rights agreements were worked out stating how much water each state (and Mexico) could remove from rivers in the Colorado River drainage basin. Hydrologists estimated that the average annual flow of the Colorado River was about 20 cubic kilometers of water per year, and the agreements were based on this amount. The hydrologists, however, had measured stream flow during a period of unusually wet years, and the real value for the average flow of the river is probably more like 16 cubic kilometers per year. This means that there is not nearly enough water for everyone who wants it, and in most years the Colorado River is completely dry by the time it reaches its mouth in the Gulf of California.

Fresh water is a renewable natural resource. But since it is available in limited quantities, conflicts can arise between those vying for its use. The Colorado River does not carry enough water for all of the human demands placed on it, let alone the needs of area plants and wildlife. How can this problem be solved? Part of the answer will lie in having a better understanding of the science of water resources.

Objectives

After studying this chapter and completing the exercises, you should be able to do each of the following tasks, using supporting terms and principles as necessary.

1. Describe the various steps of the hydrologic cycle.
2. Describe stream drainage networks.
3. Explain how water in streams moves weathered materials.
4. Describe landforms associated with streams such as deltas, alluvial fans, and features on floodplains of meandering streams.
5. Describe how water flows underground in aquifers.
6. Explain how caverns form, and describe landscapes formed by erosion of limestone.
7. Describe surface water and groundwater as natural resources, and explain how these resources can be used wisely or foolishly.

Vocabulary Terms

You should be able to define or describe each of these terms in a complete sentence or paragraph.

1. alluvial fan
2. aquifer
3. artificial levee
4. bar
5. base level
6. bed load
7. braided stream
8. cut bank
9. delta
10. discharge
11. dissolved load
12. divide
13. drainage basin
14. drainage pattern
15. dripstone
16. flood plain
17. geothermal
18. geyser
19. groundwater
20. hot spring
21. hydrologic cycle
22. infiltration
23. lag time
24. laminar
25. landform
26. load
27. meander
28. natural levee
29. oxbow lake
30. permeability
31. point bar
32. porosity
33. runoff
34. sinkhole
35. spring
36. stalactite
37. stalagmite
38. stream profile
39. suspended load
40. transpiration
41. tributary
42. turbulence
43. water table
44. well

9.1. The Hydrologic Cycle

Earth is a watery world, with about 1.36 billion cubic kilometers of water in oceans, glaciers, streams, lakes, groundwater, and the atmosphere. As illustrated in Figure 9.1, most of Earth's water—about 97 percent—is stored in the oceans. Most of the remaining water is stored in glaciers such as the enormous ice sheets that cover Antarctica and Greenland. The amount of water present in familiar places such as

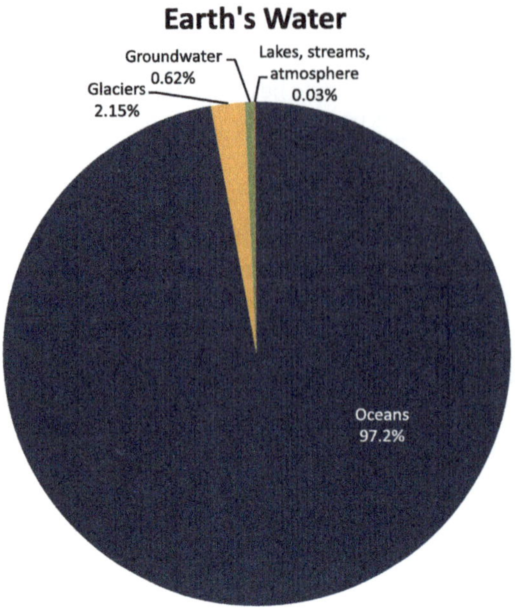

Figure 9.1. Percentages of Earth's near-surface water. The amount of water in lakes, streams, and the atmosphere is a sliver too small to see clearly.

streams, lakes, and Earth's atmosphere, is miniscule compared to the amount of water stored in the oceans, ice, and under Earth's surface as groundwater[1].

Water in Earth's hydrosphere is constantly moving. In some places, this movement is obvious—water moving in a stream, waves in the ocean, or precipitation falling from the sky. In other places it is not quite so readily apparent, such as the slow movement of a glacier, groundwater seeping between mineral grains, or water vapor moving through Earth's atmosphere (which occurs even on cloudless days). Figure 9.2 illustrates Earth's *hydrologic cycle*, sometimes referred to as the water cycle. The hydrologic cycle is the continuous circulation of Earth's water supply, characterized by the movement of water between the oceans, atmosphere, and the continents.

The hydrologic cycle is powered by radiant energy from the sun. This energy evaporates water from the oceans, other bodies of water, and Earth's land surface. The evaporated water rises into the atmosphere to become atmospheric moisture, as indicated by the rising arrows in Figure 9.2. Evaporation is an example of a phase change. Water molecules that gain enough energy—in this case from sunlight—escape from the liquid state to become gaseous water vapor. Water vapor, like other major gases in the atmosphere, is invisible. When cooled, water vapor condenses to form microscopic liquid water droplets, visible as clouds. The process of condensation is the reverse of evaporation: water molecules lose energy by cooling, slow down, and start sticking to each other to form microscopic water droplets[2]. Eventually, these water droplets merge together to form raindrops and these fall to Earth's surface as precipitation. Precipitation is indicated in Figure 9.2 by the arrows going down to the land and sea.

Most precipitation falls on the oceans. When precipitation falls on Earth's land surface, several things happen. Some of the water seeps into the soil or rock through cracks and pore spaces, a process known as *infiltration*. Much of the water that infiltrates is captured by plant roots and released back into the atmosphere from the plants' leaves through a process called *transpiration*. Transpiration is related to the process of photosynthesis. Water that avoids being taken up by plant roots infil-

1 Earth's mantle also contains a tremendous amount of water locked in the structure of minerals, but we won't consider that water in this book.

2 Or ice crystals when the temperature is low enough.

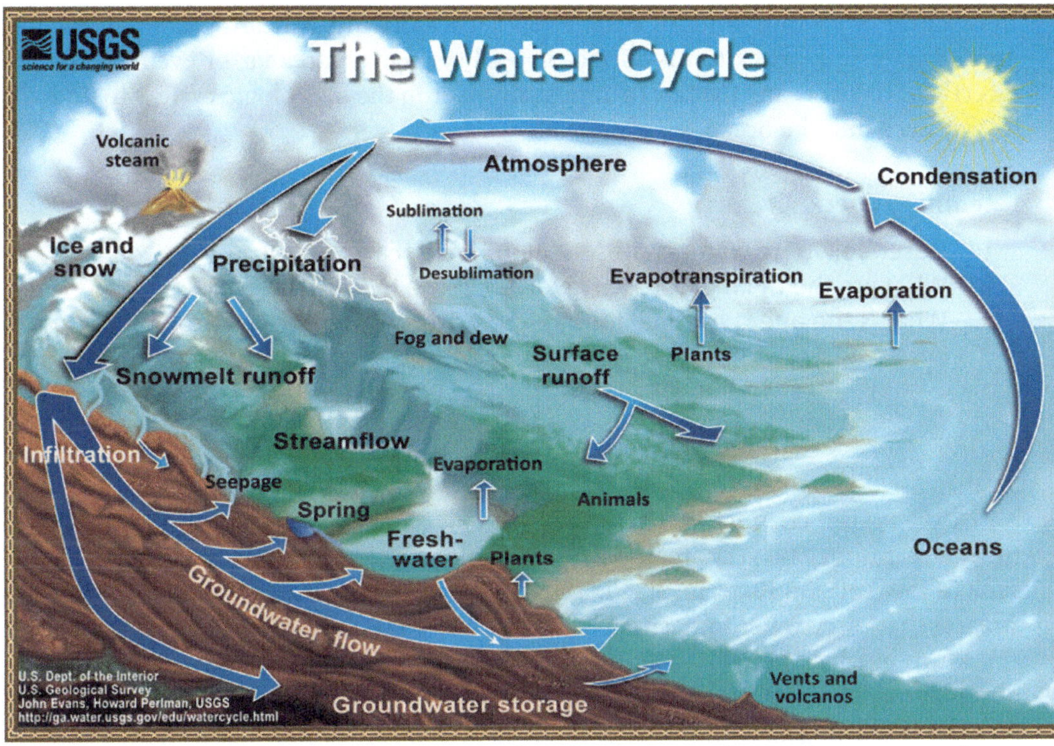

Figure 9.2. The hydrologic cycle is the ongoing natural circulation of water through Earth systems – the lithosphere, hydrosphere, atmosphere, and biosphere. The hydrologic cycle is driven by energy from the sun, which causes water to evaporate and provides energy for photosynthesis.

trates deeper to become part of groundwater. Plants are very efficient at capturing water, and in many areas the amount of water returned to the atmosphere through transpiration exceeds the amount of water that becomes groundwater.

There are several factors that influence how much water infiltrates the soil. One of these is vegetation. Soil with vegetation tends to have more pore spaces, allowing for infiltration. But denser vegetation results in a higher rate of transpiration, which returns water vapor back to the atmosphere. If soil becomes compacted by animals, human activities, or other processes, not as much water will infiltrate. The slope of the land surface also affects the rate of infiltration; water tends to flow down steep hillsides without infiltrating, while in flat areas the water sits in puddles, allowing more time for infiltration.

Water that doesn't infiltrate, evaporate, or transpire becomes *runoff*—water that flows downslope on Earth's surface. Often at the beginning of a rainstorm, the first water to fall soaks into the soil. Once the surface soil is saturated, water begins to run over the surface. Runoff eventually flows into streams or lakes and from there may eventually infiltrate into Earth's surface or evaporate.

Water that infiltrates beneath the upper layers of Earth's surface becomes *groundwater*. Because groundwater moves through cracks and pores, it moves quite slowly. Depending on the situation, water may remain underground for a short period of time or for thousands of years. Eventually, as it flows beneath Earth's surface groundwater seeps back to the surface to feed springs or streams, or even to flow directly into the ocean. The fate of almost all water in the hydrologic cycle is to find its way back into the ocean.

Learning Check 9.1

1. Describe where water is present at and near Earth's surface.
2. What are the main processes involved in the hydrologic cycle?
3. What is the energy source for moving water through the hydrologic cycle?

9.2. Streams

A *stream* is a channel on Earth's surface that carries water at least part of the time. Streams may be very small—carrying a mere trickle of water—or enormous, such as the Amazon River in Brazil. They may flow continuously or may be desert streams that carry water for a short time once every few years.

Much of what you will learn in this chapter has to do with features that can be observed on Earth's surface. These features are called *landforms*. Landforms can be described with general words such as *hill* or *valley*, or with words with specific definitions such as *oxbow lake* or *point bar* (discussed in Section 9.3).

9.2.1. Drainage Networks

A stream is always part of a larger network of streams. The land area that contributes water to a stream is called a *drainage basin*. The drainage basin of a small stream might only cover an area of a few acres[3], while the drainage basins of the largest rivers, such as the Mississippi and Amazon, cover over one million square miles.

A *divide* is an imaginary line that separates one drainage basin from another. Divides usually run along ridge lines in hills or mountains, though they can run through plains as well. Some divides, called *continental divides*, run the length of entire continents. Figure 9.3 is a map of North America showing major drainage basins and divides. Precipitation falling on the west side of the North American continental divide in the Rocky Mountains in the United States flows to the Pacific Ocean; water flowing on the east side of the continental divide runs to the Gulf of Mexico.

Within a drainage basin, the network of streams makes a *drainage pattern* that is determined by the underlying rock types and structural features such as folds and faults. Three typical drainage patterns are illustrated in Figure 9.4. In areas with flat-lying or uniform rocks, the streams often develop a dendritic[4] drainage

3 There are 640 acres in one square mile.

4 *Dendritic* means tree-like.

Figure 9.3. Major North American drainage basins and divides. The boundary between any two basins is a divide; a continental divide separates basins that drain into different oceans. The Great Basin is an arid region where streams end before reaching the sea.

pattern. Dendritic drainage patterns look sort of like the branches of a tree. If the landscape is dominated by a central high point, such as a volcano, streams will develop a radial drainage pattern. A third common drainage pattern is the trellis drainage pattern. These form in areas where there are long, parallel ridges, such as where there are parallel lines of anticlines and synclines.

Within a large drainage basin, there are a large number of small streams and one primary stream that exits the drainage basin by entering another body of water. A smaller stream that enters a larger stream is called a *tributary*. As examples, the Missouri River and Ohio River are tributaries of the Mississippi River.

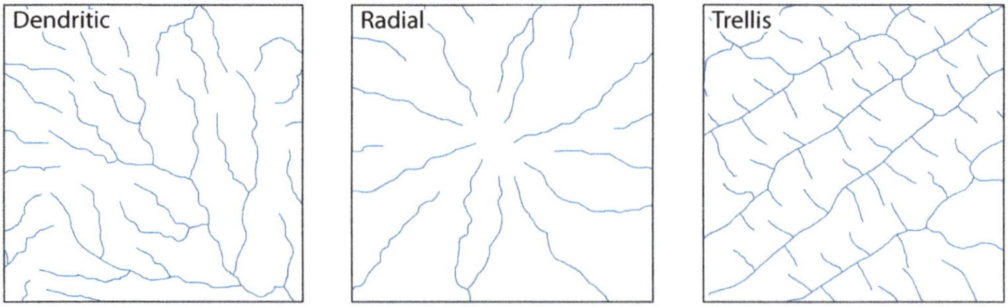

Figure 9.4. Three typical drainage patterns formed by streams. Dendritic drainage patterns are common in areas where the bedrock has a uniform resistance to weathering. Radial patterns form around single high points, such as the volcano on the cover of this book. Trellis patterns occur in areas with long, parallel ridges, such as the region shown in Figure 6.37.

9.2.2. Streamflow in Channels

Water flows downhill because it is attracted downward by Earth's gravity. The flow of water downhill is resisted by friction with the edges and bottom of a stream's channel. This friction, along with irregularities in the shape of the channel, causes water in most stream channels to flow with an erratic, swirling motion called *turbulence*. Turbulence, illustrated in Figure 9.5, causes the water to flow more slowly than it otherwise would. *Laminar* (non-turbulent) flow, in which water molecules flow in straight-line paths parallel to the channel, is much less common and only occurs in smooth-sided channels.

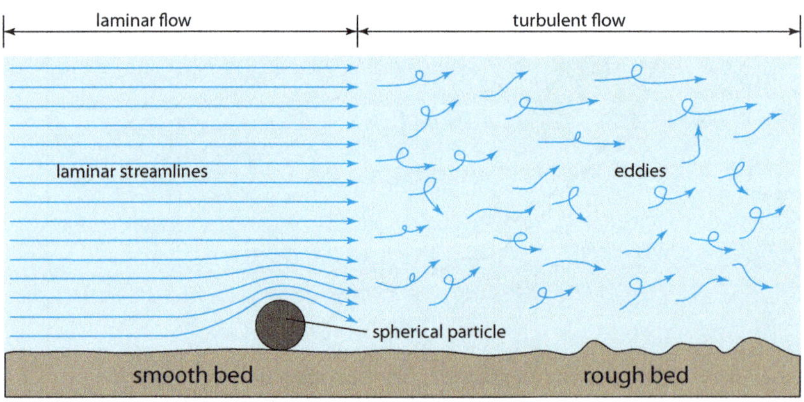

9.5 Laminar (smooth) flow and turbulent flow in a stream. Most streamflow in nature is turbulent, even in streams that appear to be gentle.

The rate at which water flows through a stream is called *discharge*. A stream's discharge is measured in units of volume of water per unit time, such as cubic meters per second (m^3/s) or cubic feet per second (ft^3/s). Table 9.1 lists the ten largest rivers in the world in order according to discharge. The discharge for any stream is variable, depending on factors such as the amount rainfall and snowmelt in the river basin.

When precipitation falls in an area, time is required for the runoff to flow over the surface into smaller streams. Additional time is required for the runoff in smaller streams to reach the larger streams. The amount of time between a rainstorm and the peak discharge for a stream is known as *lag time*, illustrated by the graph of rainfall and discharge data in Figure 9.6. The lag time in small drainage basins

Rank	River	Country	Average Discharge (m^3/s)
1	Amazon	Brazil	212,000
2	Congo	Democratic Republic of Congo	40,000
3	Yangtze	China	22,000
4	Brahmaputra	Bangladesh	20,000
5	Ganges	India	19,000
6	Yenisei	Russia	17,000
7	Mississippi	United States of America	17,000
8	Orinoco	Venezuela	17,000
9	Lena	Russia	16,000
10	Parana	Argentina	15,000

Table 9.1. Discharge of Earth's largest rivers.

is usually short—minutes or hours—since it does not take much time for water to gather from the land surface into the stream. In large drainage basins, the lag time might be several days, since it takes much more time for water to flow through all of the stream's tributaries and make it to the main stream.

A stream's velocity depends on the stream's gradient, the size and shape of its channel, and its discharge. The velocity is usually greatest near the center of the channel, where the effects of friction with the stream's edges and base are minimal. One of the most important factors determining a stream's velocity is the amount of friction the water has with its bed and banks. A deep, wide stream with a smooth bed composed of sand flows with less resistance from its bed than a steep but rocky mountain stream. The mountain stream in Figure 9.7 may look like it is flowing rapidly, but the water in this stream has a tremendous amount of turbulence, with water flowing up, down, sideways, and even backwards at times. On the other hand, the water in the wide river shown in Figure 9.7 moves along steadily at a greater average velocity than the water in the mountain

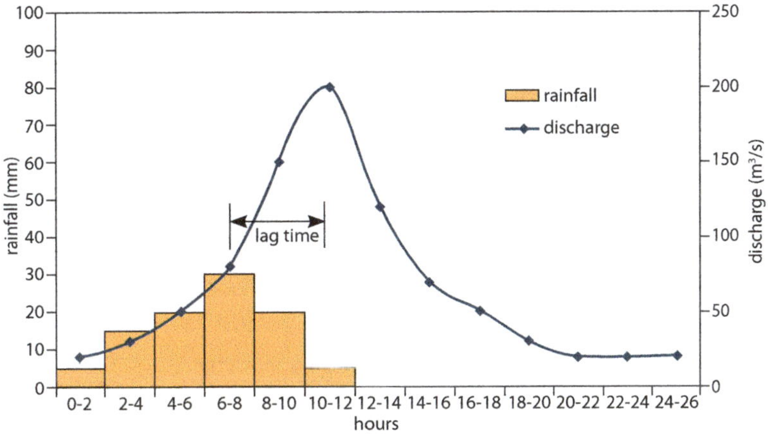

Figure 9.6. A graph showing the relationship between a rainfall event (yellow bar graph) and stream discharge (blue curve). Lag time is the gap between peak precipitation and peak streamflow.

Figure 9.7. The turbulent water in a mountain stream may appear to have a high velocity, but water in a large, deep stream with a low gradient actually has the higher average velocity.

stream. This situation is like Aesop's fable of the Tortoise and the Hare: water in a mountain stream has bursts of speed, but spends much of its time in slow-moving pools. Water in the broad stream moves along at a steady pace and wins the race.

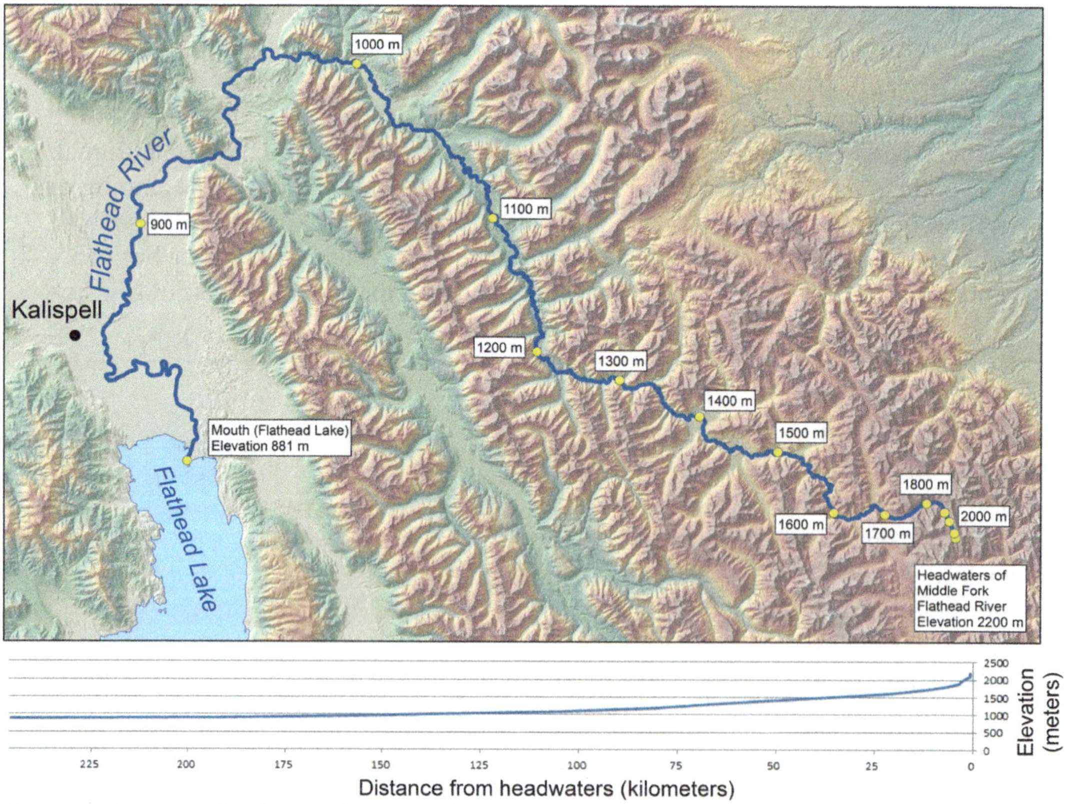

Figure 9.8. Stream profile of the Flathead River from its head in the Rocky Mountains to its mouth at Flathead Lake, 245 kilometers downstream. Stream discharge increases downstream as more tributaries flow into the river.

9.2.3. Stream Profiles and Base Level

A *stream profile* is a graph of the slope of a stream from its head to its mouth. Typically, a stream has a steeper gradient near its head and a gentler gradient near its mouth. The gradient of the Flathead River in the Rocky Mountains is 160 meters per kilometer near its source, but less than 2 meters per kilometer near its mouth on Flathead Lake, as illustrated by the map in Figure 9.8. Below the map is the stream profile, showing the elevation of the river from the source at the right to the mouth at the left. Some rivers on plains have gradients of only a few centimeters per kilometer.

As a stream flows toward its mouth, it gains water from tributaries and from groundwater that flows into the stream from beneath. To hold all this water, the channel becomes both deeper and wider. The larger channel reduces the amount of friction between the water and the streambed. As a result, the water flows faster in the deep, wide stream than in the smaller streams found at the headwaters.

A stream cannot erode to a level deeper than its mouth. The elevation of a stream's mouth is called its *base level*. Base level is a key concept for understanding whether streams will erode deeper into bedrock, deposit sediments to raise their level, or maintain an equilibrium where there is no net erosion or deposition. If the elevation of a stream's mouth changes, it will eventually have an effect on the entire length of the stream, even if the stream is hundreds or thousands of kilometers long. Figure 9.9 is a simple illustration of the valley of the Mississippi River in northern Louisiana. Twenty thousand years ago, during the most recent ice age, sea level was about 120 meters (400 feet) lower than it is today. This is because several million cubic kilometers of water now in the ocean was stored in glaciers in polar regions during the ice age. Because the base level of the river was 120 meters lower, the river eroded a deep valley. When the ice ages ended, sea level rose to its present level, and the Mississippi River responded by filling the valley with sediments.

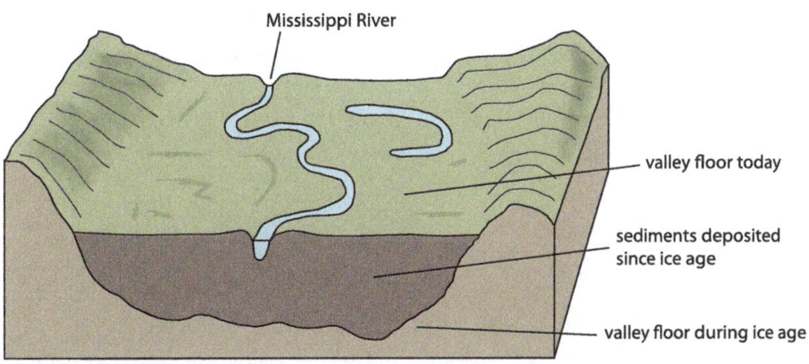

Figure 9.9. When sea level was lower during the ice ages, the base level of the Mississippi River was also lower, which caused the river to erode a deeper valley. With a higher sea level since the ice ages, the valley has filled in with sediment.

9.2.4. Erosion and Transport of Sediments by Streams

Most of the sediment carried by streams is actually the result of weathering of the landscape on slopes above the stream rather than actual erosion by the stream.

Ion	Average Concentration (mg/L)
calcium (Ca^{2+})	14.7
magnesium (Mg^{2+})	3.7
sodium (Na^+)	7.2
potassium (K^+)	1.4
chloride (Cl^-)	8.3
sulfate (SO_4^{2-})	11.5
bicarbonate (HCO_3^-)	53.0

Table 9.2. Ions commonly carried as dissolved load in streams.

This weathered material is carried downslope by overland flow or in rills and gullies, primarily during rainstorms or when snow melts. Loose material is also carried downhill by landslides, mudflows, and the slow downslope creep of soil.

Moving water picks up loose particles from the streambed. Faster-flowing water moves larger particles: a slow stream might be able only to erode clay, silt, and sand from its bed, while a high-gradient stream at flood stage might be able to move large boulders. Streambed erosion of solid bedrock occurs when gravel particles moving in the water strike the streambed, causing the bedrock to fracture. The water current then picks up these loosened particles and carries them downstream.

The amount of sediment carried by a stream is called the stream's *load*. A stream's load is proportional to its discharge; a large river normally carries a much

Figure 9.10. After a hurricane in 2011, the Connecticut River flooded and carried a large suspended load, seen in this Landsat image as a muddy sediment plume entering Long Island Sound.

greater load than a small stream. There are three types of load: dissolved load, suspended load, and bed load.

Dissolved load is the material a stream carries as dissolved substances. For example, chemical weathering causes the dissolution (dissolving) of limestone, forming calcium ions (Ca^{2+}) and bicarbonate ions (HCO_3^-). Once dissolved in stream water, these ions are carried off by the water. The most common ions carried as dissolved load in streams are listed in Table 9.2. It is estimated that streams carry about 4 billion tons of dissolved load into the oceans each year.

A stream's *suspended load* is made up of small particles, such as clay, silt, or fine sand. These particles are held "in suspension" by the turbulence of a stream's water. The suspended load is visible as a cloud of sediment in the water. In other words, suspended load is what makes water look muddy, as shown in Figure 9.10. The amount of suspended load a stream carries is related to the velocity and turbulence of the water and the size of the sediment grains.

Bed load is composed of sediments that are too coarse to be carried in suspension and are moved along the bottom of the stream. Sand and gravel carried by bed load moves by rolling, sliding, or bouncing along the stream bed. Streams carry much more bed load at flood stage than when water is low, and the bed load is capable of causing significant amounts of erosion.

9.2.5. Stream Deposition

If you mix gravel, sand, silt, and clay with water in a jar, shake the jar, and then let the sediments settle, the sediments will form layers, as shown in Figure 9.11. Grains of gravel will sink to the bottom very quickly, forming the bottom layer. Sand grains will settle to the bottom next, followed by silt and clay. The clay sometimes takes over a day to settle. If gravity were the only force acting on these sediment grains, they would all settle to the bottom at the same rate. However, there is an upward force acting on the grains as they fall through the water due to friction between the grains and water. This causes small particles, such as silt and clay, to fall through the water much more slowly than coarser particles.

Figure 9.11. The sediments and water in this jar were shaken and then allowed to settle. Coarse sediments settle in water more rapidly than fine sediments. Clay particles can take many hours to settle in a small jar, and a much longer time in lakes and the ocean.

The same process happens in streams. Just as lightly shaking a jar of sediments might not move

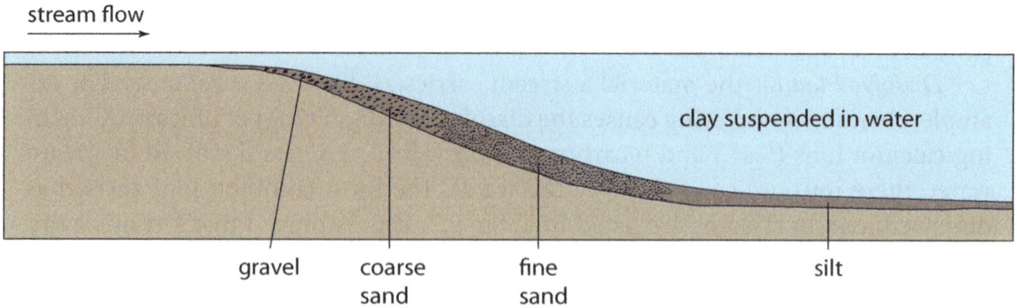

Figure 9.12. As a stream enters a lake, coarse sediments such as gravel are deposited closest to the shoreline. Fine sediments such as silt and clay are carried further out into the lake.

gravel pieces, so a slowly flowing stream doesn't move gravel. When water velocity decreases—which is what happens when a stream enters a lake or the ocean—the coarser grains are deposited first, near the mouth of the stream, followed by finer grains farther out in the body of water. The distribution of sediment grain sizes as a stream enters a lake is illustrated in Figure 9.12.

Figure 9.13. A gravel bar along a stream. Bars can also be present in the middle of the stream channel, forming islands when the water is low.

In a typical stream, water moves rapidly enough to transport clay, silt, and fine sand continuously. Gravel and coarse sand, on the other hand, are transported only at higher water velocities, such as those present during floods. When the flood subsides, gravel and sand are deposited temporarily in the stream channel as gravel bars or sand bars. A *bar* is a temporary deposit of sediment within a stream channel. Figure 9.13 shows a gravel bar in a stream.

Figure 9.14. Alluvial fans form where streams exit mountains and dump their bed load in a cone-shaped landform.

When a stream exits a narrow mountain canyon and flows out onto a plain or valley, the stream's velocity decreases quickly and sediments are dumped in a cone-shaped or fan-shaped deposit called an *alluvial fan*. Alluvial fans, such as the one shown in Figure 9.14, are especially common along mountain ranges in desert areas, where they are composed mostly of gravel and sand. Just as when a stream enters a body of water, the coarsest sediments in an alluvial fan are usually at the top of the fan, with finer sediments deposited further out on the fan.

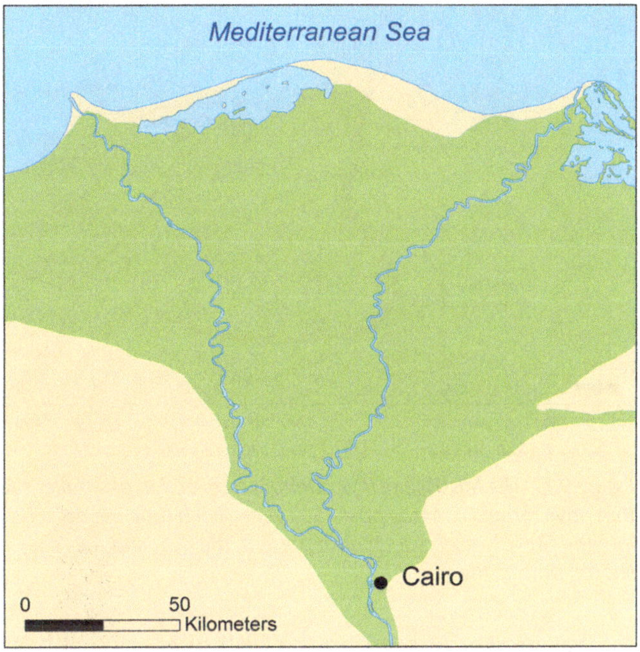

Figure 9.15. The Nile delta in Egypt has a triangular shape.

A *delta* is an accumulation of sediments formed where a stream enters a lake or ocean. Some deltas, such as the one at the mouth of the Nile River in Egypt, illustrated in Figure 9.15, are triangular in shape. The word delta comes from the Greek letter *delta* (Δ, the *d* in the Greek alphabet) because it is also triangular in shape. Many other deltas, however, are not triangular. A well-known example is the delta of the Mississippi River in the Gulf of Mexico, illustrated in Figure 9.16. The shape of the delta is determined partly by the shape of the shoreline, but the most important factor is the nature of the water currents that deposit the sediments. The strongest currents around the Mississippi River delta are those of the river itself, and the delta grows farther out into the Gulf over time. The sediments of the Nile delta, on the other hand, are reworked much more strongly by wind-driven waves after being deposited at the mouths of the river. This causes the sediments to spread out along beaches for hundreds of kilometers along the coast of the Mediterranean Sea.

Figure 9.16. The outermost extension of the Mississippi delta in Louisiana is called a bird-foot delta.

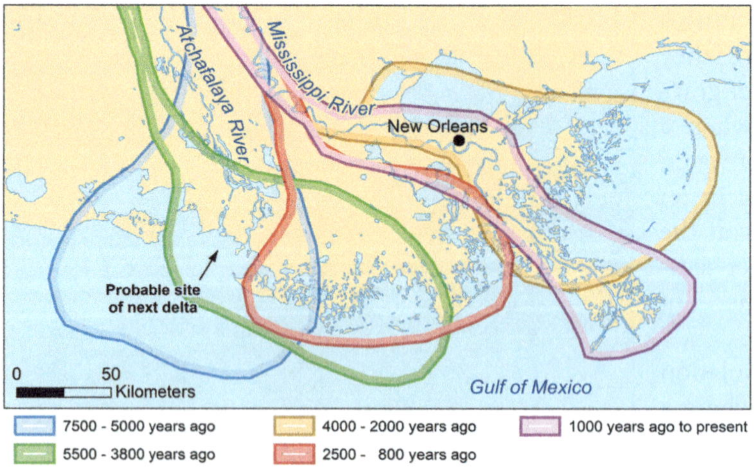

Figure 9.17. The Mississippi River Delta is a collection of smaller subdeltas that have formed over the past 7500 years. Each time the delta builds far out into the Gulf of Mexico, such as at present, the river changes course to find a shorter path to the sea.

The Mississippi River delta is the most-studied delta in the world. Geologists have unraveled much of its past development, summarized by the regions shown on the map in Figure 9.17. The Mississippi River has actually built five deltas in the past 7,500 years, each of which grew farther and farther out into the Gulf of Mexico. Each time a delta has grown far out into the Gulf of Mexico, it has gotten to a point where the river can find an easier path to the sea. When this happens, the river undergoes a major change of course, leading to the creation of a new delta. Each delta illustrated in Figure 9.17 has lasted for about 1,000–2,000 years. At present, the Mississippi River has built out into the Gulf of Mexico about as far as it can, and the shorter, more natural path for it to take would be to follow the present course of the Atchafalaya River to the west. Engineers have spent hundreds of millions of dollars on dams, levees, and other control structures to keep the Mississippi in its present course because the economies of the cities of New Orleans and Baton Rouge are dependent on shipping along the present river.

Learning Check 9.2

1. Draw sketches of three different stream drainage patterns and explain how each one forms.
2. Explain the cause of turbulence in streams.
3. Explain why peak stream discharge often occurs some time after a rainstorm rather than during the storm.
4. Explain why water in large, deep streams often moves more rapidly than water in smaller, rock-filled mountain streams.
5. Why can't a stream erode to a deeper level than the stream's base level?
6. Describe the three ways in which a stream carries its load.
7. Describe the factors that influence the shape of the delta at the mouth of a river.

9.3. Stream Landforms

Most streams flow in valleys that are proportional in size to the size of the river that formed them. The valley of a large river is usually wider than the valley of small tributaries that flow into it. Some streams flow in narrow valleys that are cut directly into bedrock, while other streams flow in broad valleys that are underlain by sediments.

9.3.1. Stream Valleys

In mountainous areas, streams are typically in narrow valleys with steep sides. These mountain streams actively erode deeper into the bedrock and have steep gradients. In places where the bedrock is more resistant to erosion, rapids or waterfalls form, such as the waterfall shown in Figure 9.18.

Figure 9.18. Many waterfalls, such as Lower Falls in Yellowstone National Park, occur where a river flows over rocks that are more resistant to erosion than rocks further downstream.

In wide valleys, streams are no longer doing much downward erosion. Instead, streams in wide valleys tend to wander from one side of the valley to another over time, and most erosion occurs at the sides of the valley.

Stream channels in wide valleys almost never follow a straight line path.

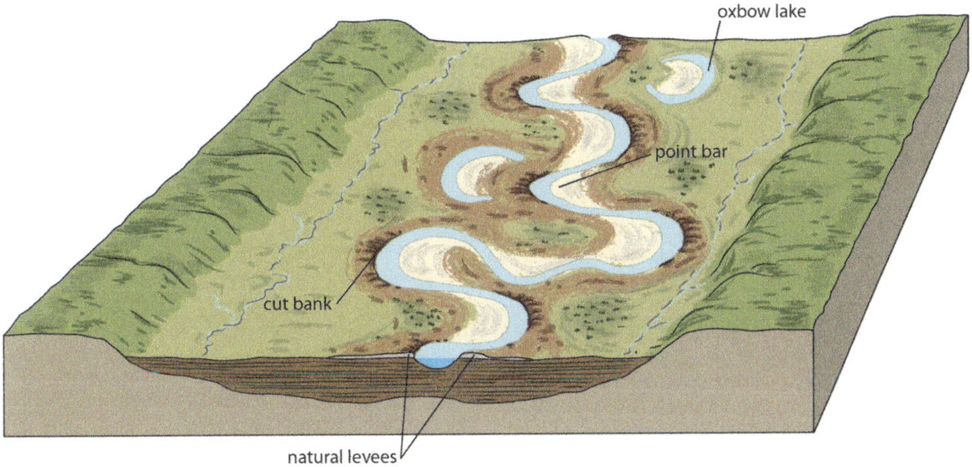

Figure 9.19. Landforms present along a meandering stream.

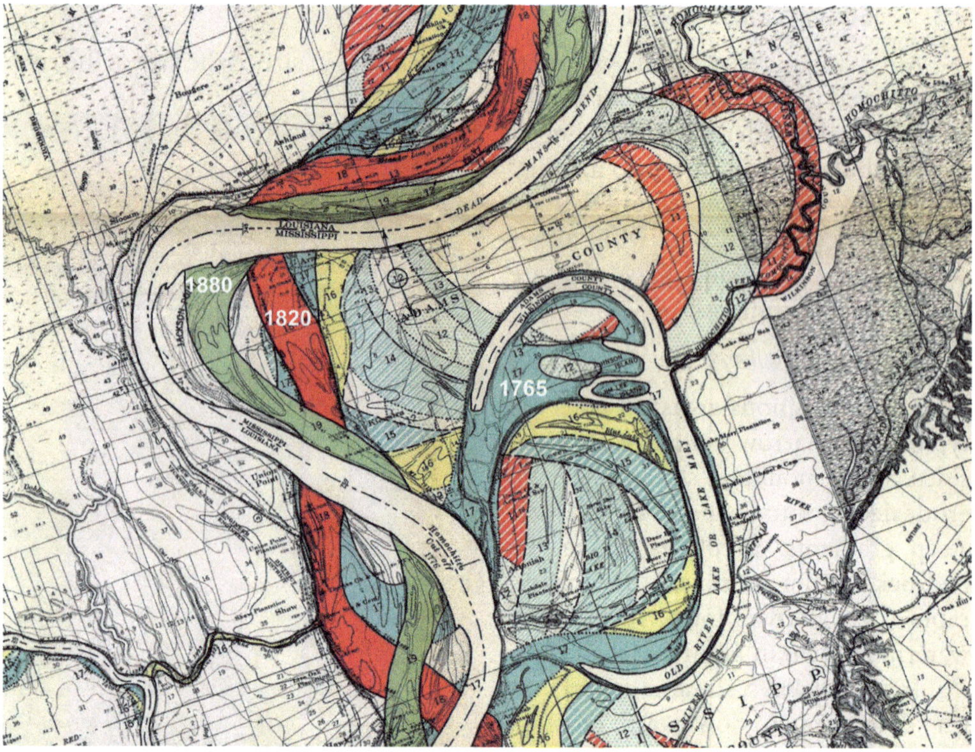

Figure 9.20. Meanders constantly move around on a floodplain. This map shows a geologist's interpretation of approximately 1000 years of meander history along a portion of the Mississippi River.

Many streams, especially those that carry significant amounts of silt and clay, have channels with looplike bends called *meanders*. As illustrated in Figure 9.19, the outer part of a meander loop is called a *cut bank* and the inner part is called a *point bar*. As water flows around a meander loop, the current flows fastest along the cut bank, leading to erosion of the material along the outside of the curve. At the same time, sediments are deposited on the point bar, and it grows larger. The result is that the paths of meander loops continuously change, as erosion and deposition occur

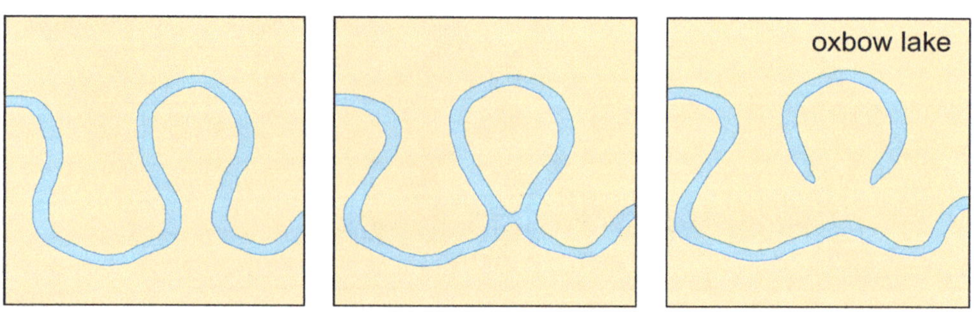

Figure 9.21. As erosion occurs on the outer edge of the curves of stream meanders, the necks of meanders become narrower and eventually are eroded through, resulting in the formation of an oxbow lake.

Figure 9.22. Arial photograph of a braided stream—the Waimakariri River in New Zealand.. The channels change their pattern within the gravely floodplain on a regular basis.

at different places. As Figure 9.20 illustrates, the Mississippi River is a meandering river for much of its length with a complex history of channel changes. The Mississippi does not change nearly as much now as it has in the past due to structures erected by engineers to prevent channels from moving.

Over time, erosion makes the "neck" of a meander loop narrower. Eventually, the stream breaks through the neck, forming a shorter path, as illustrated in Figure 9.21. The former bend is then abandoned and becomes an *oxbow lake*, a curved lake along a meandering river. Over a longer period of time, oxbow lakes become filled with silt and clay deposited when the stream floods. Eventually, oxbow lakes disappear.

Not all streams are meandering streams. Some streams, especially those that carry large amounts of sand or gravel, form "braided" channels rather than meanders. A *braided stream*, such as the one shown in Figure 9.22, has numerous small channels separated by small islands. These channels and islands are usually quite temporary, with the pattern of channels changing from year to year. Braided streams are common in glacial valleys in mountainous areas, and in sandy plains.

9.3.2. Floods and Floodplains

The flat area in a stream valley that is subject to occasional flooding is called the *floodplain* (see Figure 9.19 again). A flood is an event where stream discharge exceeds the capacity of the channel to hold water, as shown by the satellite im-

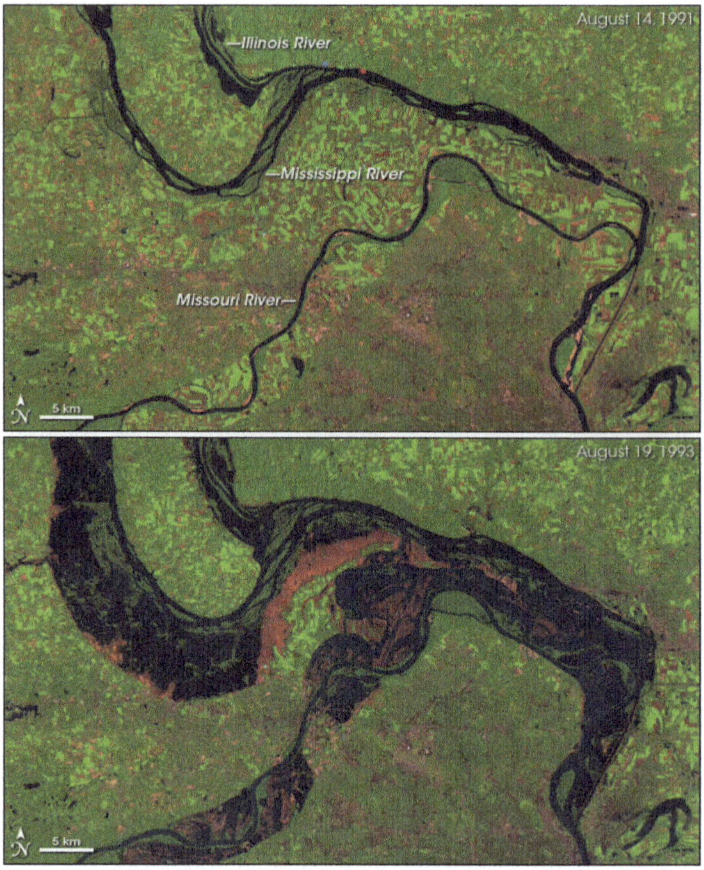

Figure 9.23. Landsat satellite image of the 1993 flood of the Mississippi River and its tributaries. In the top image, taken before the flood, the rivers are within their normal channels. In the bottom images, the rivers have overflowed to cover their floodplains. This flood lasted for about six months.

ages in Figure 9.23. Floods may be caused by weather events, such as heavy rainfall or rapid snowmelt. Rainfall from intense thunderstorms causes small stream channels to fill and overflow quickly, leading to local floods known as flash floods, such as the one shown in Figure 9.24. It usually takes excessive rainfall over a larger area, such as from hurricanes or other large storms, to cause larger rivers to flood.

Floods may also be caused or increased in size or frequency by human activities, such as dam failures or growth of urban areas. The worst dam breakage disaster in United States history occurred in 1889, when 2,200 people were killed due to a dam that failed after heavy rains, flooding Johnstown, Pennsylvania. It is estimated that the amount of flow at the peak of the flood equaled the average discharge of the Mississippi River for a short time.

The paving of large areas in cities causes an increase in flooding by reducing rainwater infiltration into the Earth, as well as by increasing the rate at which water can run off the land surface in heavy downpours.

Figure 9.24. Flash floods rise and fall quickly, and occur after heavy rainfalls, especially in arid regions.

These two factors increase the amount of water that flows into small streams and reduces the lag time between rainfall and peak stream discharge. In many communities, new parking lots must now have rainwater retention basins built next to them to collect water. The contained water is then released at a slower rate into storm sewers or streams.

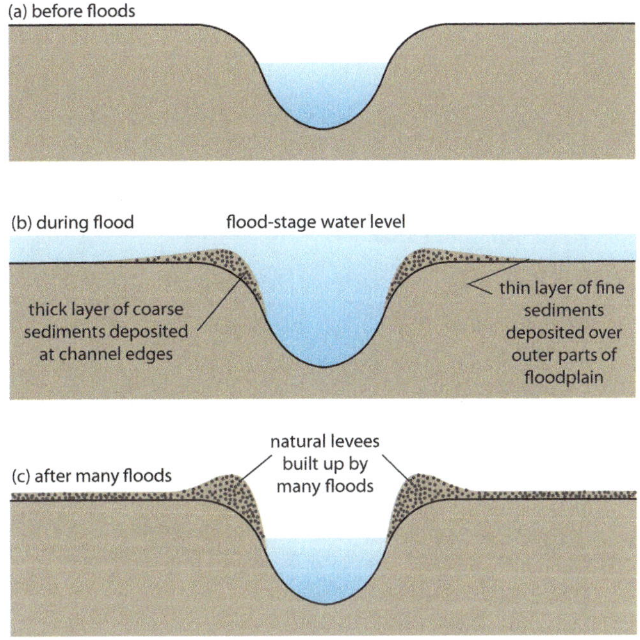

Figure 9.25. The development of a natural levee along a stream channel. Natural levees along the Mississippi River stand as high as 6 meters (20 feet) above the floodplain.

Floodwater is usually rich in silt and clay. This silt and clay is held in suspension when the water is moving in the stream channel. But as floodwater spreads out over the floodplain, the water slows down, leading to deposition of sediments. The coarsest sediments, especially sand and silt, are deposited immediately adjacent to the stream channel, creating a *natural levee*, a ridge that forms along some streams and helps to keep water confined to the channel. Illustrated in Figure 9.25, the natural levee is the highest feature on many meandering river floodplains. The rest of the floodplain has deposits of clay deposited during floods and is often naturally swampy, especially in humid climates.

Every year, floods kill numerous people and do a tremendous amount of property damage. This occurs in both urban and agricultural areas. In order to minimize flood hazards, engineers build flood control structures such as *artificial levees* and dams such as those pictured in Figures 9.26 and 9.27.

Figure 9.26. An artificial levee built by engineers to keep floodwater (on the right side of the picture) out of a community.

Artificial levees are constructed of earth or concrete and are placed along rivers to prevent flooding of communities, valuable agricultural land, or industrial sites. They also help to prevent the migration of stream channels over time, which also protects developed areas and farmland.

Figure 9.27. This flood control dam near Denver was built to reduce the risk of flooding by storing large quantities of runoff after heavy rainstorms.

Dams help to prevent flooding by storing excess water and releasing it more slowly than would occur naturally. The water stored in reservoirs created by dams is also important for irrigation, drinking water supplies, generation of electricity, and recreation.

Artificial levees and dams bring many benefits to society, but there are drawbacks to these structures. Building an artificial levee along a river actually raises the level of water in the channel during a flood because excess water is no longer free to move out over the floodplain. The result of preventing flooding in one place, therefore, is sometimes to cause worse flooding someplace else. The construction of dams to control flooding traps sediments upstream from the dam. Reservoirs eventually fill up with sediments—typically over a period of centuries—greatly reducing the benefits of the dam. The water downstream of a dam has a lower sediment load, and this causes problems as well. After the construction of the Aswan Dam on the Nile River in Egypt in the 1960s, the sediment load reaching the mouths of the Nile was greatly reduced. This has led to serious beach erosion problems along the Mediterranean Sea coastline.

Increasingly, community planners now consider alternatives to building dams and levees to control floods. One alternative is to prohibit construction of new business and residential developments on floodplains. Simply allowing some of these undeveloped areas to flood in one place reduces the level of flooding at other places.

Learning Check 9.3

1. Describe how oxbow lakes form along a meandering stream.
2. Compare meandering streams and braided streams.
3. What causes floods, and what are some ways in which humans work to prevent or reduce the damage from floods?
4. Describe how natural levees form along streams.

9.4. Groundwater

Groundwater is an important natural resource, providing up to half of the drinking water supply in the United States and a similar amount in many regions of the world. Groundwater is also an important source of water for irrigation in arid and semiarid regions and for industrial use. As shown back in Figure 9.1, there is a lot more freshwater stored underground as groundwater than the amount of water in streams and lakes.

Rather than flowing in some sort of underground rivers, almost all groundwater is found in the tiny pores between sediment grains or in fractures in crystalline rocks. As shown by the hydrologic cycle, groundwater comes from precipitation that infiltrates into the ground. Some of this water stays for a time in soil, but soil generally is not saturated, or soaked, with water. However, at a certain depth under the surface, the ground *is* saturated with water. This means that all of the pore spaces in the material are filled with water rather than air. Water in this zone of saturation is called *groundwater*.

The upper surface of this water-saturated zone is called the *water table*. Above the water table, pore spaces contain both air and moisture; beneath the water table pore spaces are filled with water. There is a close relationship between streams on Earth's surface and groundwater beneath the surface. As illustrated in Figure 9.28, the water table goes up and down with the hills and valleys of Earth's surface. The flow of groundwater through pores follows curved paths from places where the water table is high to where the water table is low, such as at streams, lakes, or the ocean. As illustrated in the hydrologic cycle in Figure 9.2, water in streams does not only come from rainfall and snowmelt; stream water also comes from groundwater. This is why many small streams can continue to flow even in dry seasons when there appears to be no source for water in the streams. Rainfall and snowmelt often vary considerably

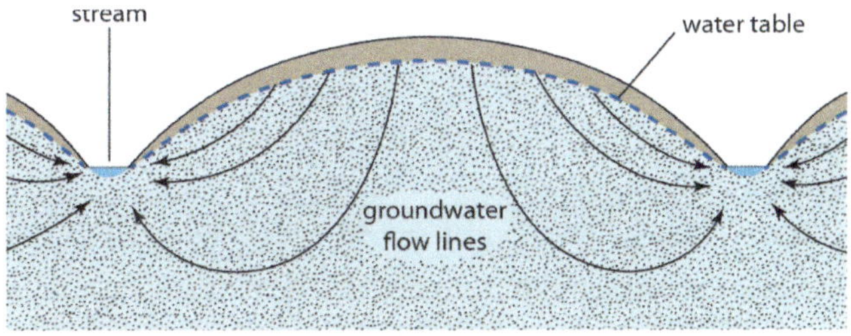

Figure 9.28. The water table is closer to Earth's surface in valleys, and further beneath the surface at higher elevations. Groundwater flows along curved paths from higher elevations to streams and other bodies of water.

from wet seasons to dry seasons, but contributions to streams from groundwater tend to be much more steady.

9.4.1. Porosity and Permeability

Some rocks and sediments can hold a large amount of water; others hold much less. *Porosity* is a measure of the percentage of pore spaces in a rock or sediment. The pore spaces may be gaps between sediment grains or fractures in rocks. Loose sediments have porosities up to 50%, while crystalline rocks like granite sometimes have porosities close to zero.

Pores must be connected to each other and large enough to allow water to flow. *Permeability* is the measure of the ability for water to move through a rock or sediment. A rock or sediment might have high porosity, but if the pores are not well-connected to each other, water cannot move through the material and the material is said to be *impermeable*. Sand and gravel, as well as highly-fractured rocks, often have high permeability. Clay and shale often have very low permeability, which makes these fine-grained materials very effective at blocking the movement of groundwater[5].

9.4.2. Aquifers

In most situations, groundwater moves very slowly, at rates of just a few centimeters per day. A layer of rock or sediments that allows the movement of groundwater is called an *aquifer*. Aquifers are composed of rocks or sediments that have both high porosity and high permeability, such as sand, gravel, silt, poorly-cemented sandstone, and other rocks if they contain enough fractures. Some aquifers are near Earth's surface, and are replenished by precipitation falling on the main surface of the aquifer. These aquifers may range from a few meters to hundreds of meters in thickness. Other aquifers can be found deeper in Earth's crust. These deep aquifers are fed by rain that falls at a place where the aquifer reaches the surface. Figure 9.29

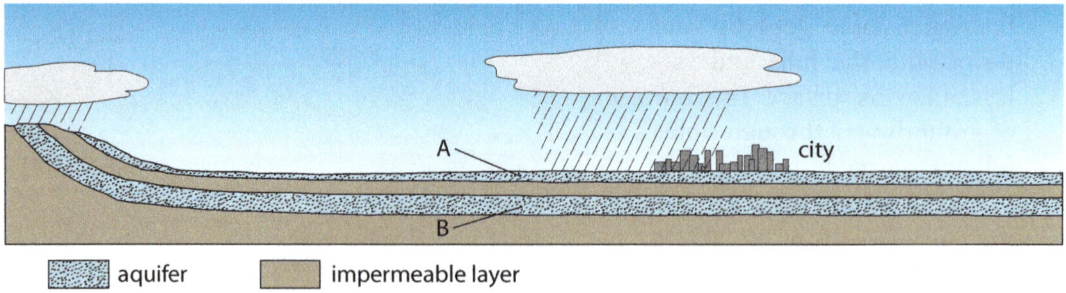

Figure 9.29. Rain that falls on the plains will enter aquifer A because it is at the surface. But surface rain cannot flow down to aquifer B because of the presence of an impermeable layer. Aquifer B can only be recharged by rain that falls in the mountains some distance away.

5 The concept of permeability also applies to the ability of petroleum to move through rocks. The controversial process of "fracking" uses explosives and chemicals deep underground to fracture rocks, greatly increasing the permeability of petroleum-bearing rocks and allowing the petroleum to flow more easily to oil wells.

illustrates near-surface and deep aquifers. The deep aquifer is separated from the surface by impermeable layers, so it cannot be replenished by rainfall on the plain where the city is. However, the deep aquifer emerges at Earth's surface at some distance from the city and is recharged by rainfall that falls there. The recharge area for some aquifers is hundreds of kilometers away from the main body of the aquifer.

Figure 9.30. The water for these springs along the Snake River in Idaho traveled over 100 kilometers through an aquifer.

Figure 9.31. A large hot spring in Yellowstone National Park. The colors are caused by microorganisms that thrive in the near-boiling temperatures of the water.

9.4.3. Springs and Wells

At places where the water table intersects the surface someplace other than a streambed, water flows from the ground as *springs*, such as those shown in Figure 9.30. If the water has flowed rapidly from great depths, or if there is magma or hot rocks at a shallow depth, water emerges from the ground as a *hot spring*, such as the hot spring in Figure 9.31. The temperature of water in hot springs ranges from human body temperature (37°C, 98.6°F) up to boiling (100°C). A *geyser*, such as the one shown in Figure 9.32, is a special type of hot spring that periodically shoots water into the air. A geyser is formed when underground chambers in hot rocks are connected to the surface through small openings. Geysers only occur in a handful of places on Earth.

In some places, heated water from beneath Earth's surface is used to heat homes and other buildings or to generate electricity. The term *geothermal* refers to the internal heat of Earth, and geothermal heat is used to heat houses and generate electricity from geothermal steam. At a geothermal power plant, such as the one

Figure 9.32. A geyser is a type of hot spring that periodically erupts, shooting steam and hot water into the air. Grand Geyser in Yellowstone National Park, pictured here, erupts to heights of up to 200 feet (60 m).

Figure 9.33. Geothermal power plants, such as this one in Iceland, use steam produced in Earth's interior to generate electricity.

pictured in Figure 9.33, steam from beneath Earth's surface is used to turn turbines to generate electricity. Geothermal energy is considered to be a renewable natural resource, and as with all natural resources, it can be used wisely or foolishly.

Significant quantities of water are removed from aquifers through the use of wells, such as the one shown in Figure 9.34. A *well* is a hole dug or drilled into an aquifer for the purpose of supplying water. Prior to the use of electric power, all wells were dug by hand. In less developed parts of the world, hand-dug wells are still an important source for water. But today, most wells are drilled by machines and use electric pumps to bring water up from the ground. In Figure 9.35, each of the circular irrigated fields has a single well in its center, with irrigation sprinklers on long pipes

Figure 9.34. Most wells require some sort of pumping to bring water up from beneath the water table. Many pumps are now powered by electricity.

mounted on wheels that rotate around the center of the circle like a hand on a clock.

9.4.4. Groundwater as a Resource

Groundwater is a renewable natural resource because water in most aquifers is constantly renewed by infiltration through the hydrologic cycle. When water is pumped out of an aquifer, the water table goes down (becomes deeper). If too much water is removed from the aquifer, shallow wells become dry, requiring them to be abandoned or drilled deeper. In order for groundwater pumping to be sustainable, the amount of water that is removed from an aquifer must be no greater than what can be replaced by infiltration elsewhere. In many areas of the world today—especially in arid regions—the amount of water pumped out of the ground far exceeds what is replaced by the hydrologic cycle.

Figure 9.35. Circles formed by center pivot irrigation systems. Each circle in this photograph has a diameter of about a half mile.

In many places, groundwater can be pumped from the ground and used for drinking water with little or no treatment to remove microorganisms or substances that would cause health problems for humans. However, groundwater can become polluted through human activities. The most common sources of groundwater pollution are sewage, landfills, industrial waste, and agricultural chemicals such as pesticides.

Learning Check 9.4

1. What is the source of virtually all groundwater?
2. Describe how groundwater flows from a hilltop to a stream.
3. What is the difference between porosity and permeability?
4. What happens, in a short time period and over a long time period, when more water is pumped from an aquifer than is resupplied by rainfall?

9.5. Caverns and Groundwater landforms

In a previous chapter we saw that acidic water causes chemical weathering at Earth's surface. Groundwater is often acidic as well and when it is, it is capable of dissolving limestone beneath Earth's surface. This leads to the formation of *caverns*[6]. A cavern is a naturally formed underground chamber most often produced by the dissolving of limestone. The largest caverns in the world have hundreds of kilometers of passageways.

Figure 9.36. Lechuguilla Cave in New Mexico has spectacular dripstone features, including these stalactites and stalagmites.

Caverns form at or just below the water table when acidic water moves through various cracks in limestone. As limestone dissolves, these fractures enlarge, sometimes creating enormous underground rooms. The largest chamber at Carlsbad Caverns in New Mexico covers an area the size of fourteen football fields.

Caverns are formed by erosion, but they also contain features formed by deposition. Water dripping from the ceiling or walls of chambers contains dissolved calcium carbonate, which forms deposits called *dripstone* on the ceiling, floor, and walls of the cavern, as shown in Figure 9.36. This calcium carbonate is dissolved from the surrounding limestone and deposited under the different conditions present in the cavern. Icicle-like dripstone features hanging from a cavern's ceiling are called *stalactites*. Features that grow up from the floor due to dripping from above are known as *stalagmites*. Stalagmites typically form directly beneath stalactites. Stalactites and stalagmites grow longer over time, and sometimes join to form *columns*.

Even though groundwater flows beneath Earth's surface, the work done by groundwater can have a profound influence on Earth's surface. If a cavern is located at a shallow depth beneath Earth's surface, the cavern's roof may collapse to form a circular depression known as a *sinkhole*, as shown in Figure 9.37. Sinkholes may form suddenly, which can be a considerable hazard for humans living in the area, or they may form slowly as the cavern roof collapses bit by bit.

6 A *cave* and a *cavern* are the same thing, though some people use the term cavern only for larger caves.

Figure 9.37. Bird's-eye view of a sinkhole that formed during a drought in central Florida. The sinkhole destroyed homes, businesses, and the community swimming pool (on the left of the image).

As groundwater and surface water continue to weather and erode large areas of limestone, the landscape may become so dominated by caverns and sinkholes that distinct landforms are created. Figure 9.38 shows an area of China where most of the limestone has been dissolved away, leaving behind isolated towers of rock.

Figure 9.38. Large areas of southern China have isolated towers left behind by the dissolving of limestone.

Learning Check 9.5

1. How do caverns form?
2. What is dripstone?
3. Explain how stalactites, stalagmites, and columns form in caverns.
4. Describe how sinkholes form in areas with caverns.

Chapter 9 Exercises

Answer each of the questions below as completely as you can. Write your responses in complete sentences unless instructed otherwise.

1. What are ways that people in the southwestern United States could more effectively make use of water from the Colorado River? Think of ways to conserve water used for farming, communities, and recreation. Also think of alternative sources of water for the region.
2. Draw a sketch of the hydrologic cycle, with labels showing where water is stored and arrows showing the movement of water through the cycle.
3. The Nile River, with a length of almost 7,000 kilometers, is the longest river in the world, but its annual discharge is only about 15% that of the Amazon River. Form a hypothesis as to why the Nile River carries so much less water than the Amazon River does.
4. Using what you know about how gravel, sand, and silt settle out in a water-filled jar, predict how sediment grain size might change on a large alluvial fan as one goes further away from the top of the fan.
5. Explain why the mouth of the Mississippi River shifts its location every few hundred years.
6. Explain how the growth of cities has led to the increase of flash floods in and around those cities.
7. Explain why it is important for groundwater resources to be conserved and sustained.
8. Speculate on how geothermal steam for electricity generation can be used in sustainable and unsustainable ways.

Appendix A

Minerals

Part 1 Minerals with Metallic Luster				
Mineral (Formula)	Color (Streak)	Hardness	Breakage Pattern	Comments and Uses
Graphite (C)	black to gray (black to gray)	1–2	microscopic flaky cleavage	Pencil lead, lubricants
Galena (PbS)	silver gray (gray to black)	2.5	perfect cubic cleavage	Very dense; an important ore of lead
Silver (Ag)	silvery white (light gray or silver)	2.5	fracture	Tarnishes to black; used for coins, jewelry, electronics
Gold (Au)	golden yellow (yellow)	2.5–3	fracture	Does not tarnish; used for jewelry, investments, dental fillings
Bornite (Cu_5FeS_4)	bronze, dark blue, purple (gray-black)	3	uneven fracture	A copper ore, often called "peacock ore" because of its purple and blue tarnish
Copper (Cu)	copper red (copper red)	3	fracture	Used in wires, pipes, coins, and many other uses
Chalcopyrite ($CuFeS_2$)	brassy or golden yellow (green-black)	3.5–4	uneven fracture	An important ore of copper; distinguish from pyrite by hardness
Chromite ($FeCr_2O_4$)	black or brown (brown to black)	5.5	uneven fracture	Ore of chromium, used in stainless steel
Pyrolusite (MnO_2)	gray or black (black)	6*	uneven fracture	May be somewhat crumbly or earthy; an important ore of manganese
Hematite (Fe_2O_3)	black or reddish brown (reddish brown)	6	uneven fracture	An important iron ore; can be black and metallic, or reddish brown and earthy
Magnetite (Fe_3O_4)	black (black)	6	curved fracture	Naturally magnetic
Pyrite (FeS_2)	light brassy yellow (greenish, brownish, black)	6.5	uneven fracture	An ore of iron, called "fools gold"

*May be as low as 2 in some samples.

Part 2 Minerals with Nonmetallic Luster				
Mineral (Formula)	Color (Streak)	Hardness	Breakage Pattern	Comments and Uses
Talc ($Mg_3Si_4O_{10}(OH)_2$)	white, light green, gray (white)	1	uneven fracture	Very soft, greasy feel
Bauxite ($Al(OH)_3$ plus iron oxides)	white, gray, pink (white, light brown)	1–3	uneven fracture	An important aluminum ore; often contains small, spherical, brownish masses in a lighter background
Sulfur (S)	yellow (yellow)	1.5–2.5	uneven fracture	May be crystalline, but is often massive or earthy
Gypsum ($CaSO_4 \cdot 2H_2O$)	white, pink, or colorless (white)	2	3 cleavage directions	Used in drywall
Muscovite Mica ($KAl_2Si_3O_{10}(OH)_2$)	colorless, light brown (white)	2–2.5	1 cleavage (thin, flexible sheets)	Colorless mica
Halite (NaCl)	colorless or white (white)	2.5–3	perfect cubic cleavage	Table salt; distinguished from calcite by cleavage
Biotite Mica (complex K, Mg, Fe, Al silicate)	black, dark green, dark brown (white)	2.5–3	1 cleavage (thin, flexible sheets)	Black mica
Barite ($BaSO_4$)	white, yellow, red, green (white)	2.5–3.5	1 cleavage, which may not be visible	Very heavy
Calcite ($CaCO_3$)	colorless or white (white)	3	3 cleavage directions, not at right angles	Hydrochloric acid causes calcite to fizz
Fluorite (CaF_2)	colorless, blue, green, purple, other colors (colorless)	4	cleaves in four directions	Glassy luster; glows under ultraviolet light
Limonite ($FeO(OH) \cdot nH_2O$)	brown or yellow (pale brown or yellow)	5–5.5	uneven fracture	Properties are variable, often appears massive or earthy
Hornblende amphibole (complex Ca, Na, Mg, Fe, Al, Ti silicate)	green to black (gray to white)	5–6	2 cleavage directions	A common mineral in granite and similar rocks; six-sided cross section
Augite Pyoxene (complex Ca, Na, Mg, Fe, Al silicate)	black (colorless)	6	2 cleavage directions, not at right angles	Square or 8-sided cross section; common in darker igneous rocks
Orthoclase feldspar ($KAlSi_3O_8$)	pink, white; rarely green or yellow (colorless)	6	2 cleavage directions at 90° angle	An important mineral in granite and similar rocks

Part 2 Minerals with Nonmetallic Luster				
Mineral (Formula)	Color (Streak)	Hardness	Breakage Pattern	Comments and Uses
Plagioclase feldspar ($NaAlSi_3O_8$ or $CaAl_2Si_2O_8$)	white or gray (colorless)	6	2 cleavage directions at 86° angle	The most abundant mineral in Earth's crust; mineral grains often have parallel lines (striations)
Olivine ($(Mg, Fe)_2SiO_4$)	green (colorless)	6.5	curved fracture	Glassy luster; common in dark igneous rocks; can be a gemstone
Quartz (SiO_2)	colorless, various colors (colorless)	7	curved fracture	Glassy luster; used in glass manufacturing, electronics
Garnet (complex Mg, Fe, Ca, Mn, Al silicate)	deep purple, red, green, black (colorless)	7.5	curved fracture	Can be a gemstone; used as an abrasive
Corundum (Al_2O_3)	colorless, green, brown, blue, red (colorless)	9	fracture	Commonly used as an industrial abrasive; rubies are red, sapphires are blue

CPSIA information can be obtained
at www.ICGtesting.com
Printed in the USA
FSOW03n1102220715
9076FS